WALKING AWAY FROM THE LEDGE

WALKING AWAY FROM THE LEDGE

a soldier's memoir

CHRIS KRUGER

MILSPACE BOOKS
an imprint of W. Brand Publishing
NASHVILLE, TENNESSEE

j.brand@wbrandpub.com
W. Brand Publishing
www.wbrandpub.com

Cover design by JuLee Brand / designchik

Walking Away from the Ledge / Chris Kruger —1st ed.

Available in Paperback, Kindle, and eBook formats.
Paperback: 979-8-89503-018-9
eBook: 979-8-89503-019-6
Library of Congress Control Number: 2025909602

CONTENTS

IN MEMORIAM

I will never forget the fallen brothers with whom I served. It was always my intent to honor them throughout this book. In order since 2003:

Gleason: Iraq, 2003

Long: Iraq, 2003

Griffin: Iraq, 2003

Brown: Iraq, 2003

Cornett: Iraq, 2006

Elizalde: Iraq, 2007

Nunez: Afghanistan, 2008

Loredo: Afghanistan, 2010

Smith: Afghanistan, 2012

Hairston: Afghanistan, 2014

Cantrell and crew: Oahu, 2017

Blizard M.: Nebraska, 2018

Blizard C.: North Carolina, 2021

Seagrave: Alaska, 2022

Dibble: Virginia, 2022

Esparza and crews: Fort Campbell, 2023

Farmer: 2023

Conklin: Tennessee, 2023

Gone but *never* forgotten: All gave some, they gave all. God bless the families that remain of all our fallen comrades. RLTW.

Someone says, "We are free!"
but freedom comes with a fee.
Of blood, sweat, and occasional tears.
Anxiety, excitement, and feeling fear.
From Soldiers who have come and went
time in service they have spent.
Under fire from big to small,
The soldier must endure it all.
Living here from day to day
far removed from any usual way.
Hope is what keeps you alive.
A goal or something toward to strive.
Minimal comforts to boost morale.
A letter from family or old pal.
The heat beats daily on your head,
sweltering to the point you wish you were dead.
To be the victor, that's the key,
and have stories for eternity.
Friends made here last forever.
To leave them in loss of battle, *never!*
We've all been through good times and bad.
Through happy hours and many sad.
But to come home to family and friends
and tell them of the tour's end.
About all the times of fear
And wishing I were never here.
This will truly be a glorious time.
To share the experience with friends of mine.
To tell the tales about Iraq
secretly hoping never to go back.

Unbeknownst to me at the time, I would deploy to the Middle
East more times than I care to remember.

PROLOGUE

It was the first week of October 2014, and we were stationed at Fort Campbell, a US Army base split between Tennessee and Kentucky. We had purchased a new, two-story house just outside of Clarksville, Tennessee, but life was not as stable as it might have seemed from the outside. I had given up my sobriety that August after learning a buddy of many years had been killed in action (KIA) in Afghanistan. On top of that, my frustrations were exponentially increasing due to professional pressure. I had just returned from New Mexico and high-altitude flight training— well, higher altitude than the hills around Fort Campbell, but there are few "high" mountains in New Mexico—when the Army, in its infinite wisdom, presented another request. It had been decided that I needed to attend and complete the Air Assault course, a physically rigorous program, even though it had been a mere eight weeks since I'd had surgery to repair an inguinal hernia.

I wasn't exactly thrilled when my first sergeant (1SG) brought it up.

"When are you going to Air Assault School?" he asked.

"I'll go when the rest of you POGs (people other than grunts, pronounced "pogue") go and graduate Ranger School," I replied with a laugh.

The 1SG didn't seem amused and stared at me blankly as if he was thinking of a comeback. I saved him the embarrassment, and even though I wasn't in the best shape of my life since the hernia surgery, I quickly told him to put me in the first class at the end of September. I didn't feel like arguing with anyone about attending the course since there was a base-wide effort to get as many soldiers Air Assault qualified as possible, if for nothing else than the historical pride of the 101st Airborne Division. Besides, maybe it would inspire the younger enlisted soldiers to attend and try something more challenging than just working on helicopters.

I had just turned thirty-five, I was fresh off surgery, and still, I would ultimately finish in the top one-third of all the physically demanding challenges that soldiers ten to fifteen years younger than me struggled with. It was pathetic and only fueled the fire of my irritation about what the Army was turning into. A weaker, coddled generation that seemed not to carry its own weight.

I was grateful when on a couple of days during the more mundane aspects of the course, I was allowed to take a half-day off. After completing some of the written exams scattered throughout the course, I was released at one o'clock in the afternoon. On one such day, I decided to swing by the liquor store to celebrate the early freedom with a couple of shots of Maker's Mark, my favorite whisky.

My wife, Genevieve (who goes by Jenny), wasn't home yet so I warmed myself up with three shots. It was just enough to get that good, lightheaded, airy buzz that I was looking for. The kind of buzz that helps you feel happy and enjoy turning up the music volume a bit more than you normally would. After I changed into some comfortable, around-the-house clothing and prepped my vest, helmet, and other required gear for the following days of instruction, my wife and daughters arrived home.

Shortly after preparing our daughters a late lunch, my wife and I had an argument that escalated into yelling. It was a nonsensical argument that centered around one of us wanting to be right. My point of view mattered more than hers on the issue, or vice versa. The type of argument that most couples have at some point in their relationship, one that gets out of hand and is only resolved when somebody gives up and both back off to let their flared tempers cool. I'd had enough and was getting ready to leave when Jenny followed me out of the house, red-faced and screaming. I wasn't having it.

"Why do you always get to walk away?" She screamed at me on the front porch.

"Stop yelling at me outside and in front of the whole neighborhood!" I shouted back as I tried to get her inside, embarrassed by the outdoor display we were providing.

Once inside, the tirade continued in front of our kids—who were four-and-a-half and two years old at the time. Then I decided to put her in a rear naked choke hold for a couple of seconds. I wanted to end the argument, and I couldn't think of another way. I would never hit her, but I wanted her to know that her behavior was unacceptable.

After I released my hold, she dashed to her phone and said, "I'm calling 911!"

"No, you're not calling the cops!" I sharply retorted.

That's when everything started.

CHAPTER ONE

"Hey! Have you thought about joining the Army?" The recruiter almost shouted over my shoulder as I perused the mall, eyeing shops I had no intention of entering. I was twenty years old, making the rounds at NorthTown Mall in Spokane, Washington, where I was born and raised on the outskirts of Spokompton in Chattaroy and Deer Park.

"Fill out this card, and I can get you on a new life path in the Army."

"I'm not interested." I tried to ignore him by not making eye contact.

"Would you fill it out anyway? I'm just trying to get rid of these cards."

"Fine." I rolled my eyes and filled out his stupid card with a fake name and phone number. Then, I went on my merry way to shop without money.

As luck would have it, I bumped into him six months later when he pulled into the parking lot where I was working events in downtown Spokane.

I approached him and told him, "Five dollars for parking."

"I'm not staying," he replied. "Are you interested in joining the Army?"

"Seriously?" I said sarcastically. Since I recognized him, I gave him my actual address and phone number this time without intending to answer or respond to calls or mail.

He returned to his car and drove off. I figured he might remember me, but I also assumed he had dozens of other teens fill out his recruiting cards and might not recognize me from any of them.

The next day, the house phone rang, and I saw a Spokane number on the caller ID. "Hello?"

"I'm looking for Chris. Is he available?" The voice on the other end asked.

I didn't know to whom I was talking but quickly remembered the day's previous interaction with the recruiter and said, "My brother's not home. Can I take a message?"

"Could you tell him the Army recruiter called and have him call me back at this number?"

"Sure," I replied unenthusiastically, then noted the phone number so I wouldn't bother answering it again.

I had never considered joining the military, in fact, I'd been opposed to it and had expressed as much to my high school guidance counselor. I was strapped for cash and not interested in college, so she had proposed, "How about the military?"

"No!" I quickly and emphatically replied.

I knew I would never join the military, but the popular action movies of the time did have me thinking about my significance as a man. *The Rock, Con Air,* and *Grosse Pointe Blank*—shoot, even the '80s hits *Rambo, Commando,* and *Predator* had all made an impression. What did all these movies have in common that resonated with me then? The military-trained badasses, those unstoppable forces of nature, who knew what to do in all situations and who were relentless until the mission was complete.

Still, joining the military was a big commitment and thinking about the future was something I didn't really do. I was twenty, and at that stage in life I just focused on the present and having fun. Not to mention, I had developed a taste for weed and alcohol and joining the military would mean giving those up.

That appetite had been sparked when I was sixteen, on my first job at Godfather's Pizza. "Can I smoke a bowl in your car?" a coworker asked.

"What is that?" I gave him a confused look.

"I want to sit in your car and smoke weed so I'm not cold." He tilted his head toward the window. "It's snowing."

"No!" I blasted.

After some back-and-forth about the whole ordeal, I finally relented and told him to leave the windows cracked so I didn't have to smell it later.

"It's not like cigarette smoke. It'll clear out before you get off. I'll leave some in your ashtray for your troubles."

"You don't have to. I don't want it," I told him.

Still, when I went to my car, I saw that he had left me a tiny little bud as a "thank you" of sorts. I left this little bitty bud sitting in my ashtray for two weeks, my curiosity growing. About six months prior, I had taken two puffs off a bowl with my cousins, but as most people will say of the first time they smoked, I didn't feel anything. *Would I feel something this time? What would it be like?* I knew one friend, maybe two, who had tried it, and they weren't addicted. It didn't seem to ruin their lives. Needless to say, I eventually gave in to temptation, and along with a buddy, we decided to light this little guy up on a Friday after school.

"Wow, this is some good stuff!" he said, leaning back in his seat. "Where did you get it?"

"A guy I work with." I watched as he smelled and squeezed it.

"It's sticky. It should smoke well." He had a pipe that he put the bud in and proceeded to light it up.

For such a small piece it seemed to last forever. We smoked on it for a good five minutes, maybe longer. Afterward, we were high as we could be. Even my other buddy, hanging out in the car with us, felt a buzz. He was angry since he had a later-than-usual basketball practice and didn't want to go feeling like that.

"If you have a cigarette after smoking weed, it increases the high," my buddy said.

So, I tried that, and sure enough, he was right.

"Have a beer, man, keep the buzz going." He handed me one.

I didn't like the beer flavor, but I liked the buzz and feeling "funny." After that, it became a normal weekend thing.

One thing led to another, and before I knew it I was being invited to parties and everyone wanted me around. I had a handful of buddies I'd hung out with since the second grade, but now my friend network expanded to more than just those and up through the grades to the senior class. I was funny with all my smack-talking at work and then hilarious with my high and drunk antics after work. If I had at least ten bucks to throw in on weed or alcohol, no one would bat an eye.

All the while, I didn't think this type of behavior was an issue because I had good grades in school—well, good enough anyway. I was always a B+ student, and partying didn't seem to affect that. My family was oblivious as long as I called in or let them know where I would be, so I continued my partying lifestyle in secret. No harm, no foul.

After graduating high school in the summer of 1997, I continued in this manner, and then finished only one semester of community college. I went to college for architecture but became fed up with mechanical drafting and the garbage math and English classes I had to take. Those two classes were so basic a sixth grader could have passed them. We only had to show up on test days, so I only had to attend one in-person class. Then, it was on to work and partying, which I no longer waited until the weekend to do.

I kept up the work-to-party lifestyle for a couple of years, which resulted in changing jobs and finally getting fired from the parking gig. Even though I had plenty of people to hang out and party with, I knew I was missing out on something more, but I didn't know what it was. Burning out of that loser lifestyle, I was a recruiter's dream.

It shouldn't have been a surprise that fate would bring me to the recruiter yet a third time. I was in between jobs with no real prospects as I walked down Division Street, across from the mall where I had first met him many months prior. I was moseying to a buddy's house to get drunk or high—maybe both—when the recruiter pulled his car into the parking lot next to where I was on the sidewalk.

He parked right next to me, nearly jumped out of the car, and said, "Fill this out."

I laughed at seeing the same guy again. I gave him accurate contact information and told him, "By the way, I do not have a brother." His puzzled look confirmed that he didn't remember me from earlier encounters or calls, and after a few moments of blank staring, I cleared my voice and said, "Never mind. Talk to you soon." He hopped back in the car and drove off.

This could be my chance to be like those movie guys, I thought. So, at the recruiting station I jumped at the opportunity to become an infantryman. After all, I thought that infantry and tanks were what the Army was.

I was going to become an unstoppable force to be reckoned with! Just the thought made me feel so much more confident, like there was nothing I couldn't achieve. You see, for far too long, I had allowed people to push me around and had even ended up in a couple of fights, which I had lost. But now, I wouldn't let that happen ever again. I was taking control.

After completing all the paperwork and testing to enlist in the Army, I was ready to go. I knew it would be for the best—or at least that's what I told myself. I quit drinking and smoking everything and I was ready for my fresh start. My family didn't want me to go, but we attended the meetings together with the recruiter and eventually agreed it was what I was supposed to do. My contract stated that I signed up for the US Army Infantry with Airborne Ranger assessment. I was so sure of my commitment that I signed up for the maximum time allowed on an initial contract: six years.

CHAPTER TWO

I got off the bus from the Atlanta airport and quickly wondered why I was sweating while standing still. I had never experienced humidity in that kind of heat. June 20, 2000, and I had never ventured too far from the Pacific Northwest. Eastern Washington is dry. I know, go figure; I bet you thought it rained incessantly but nay, my friends! That's the Seattle area. If you thought that about the rain, I bet you didn't know that the center of Washington State is a desert. True story.

It was rather chill for the first twelve hours after my arrival in Fort Benning. We got off the bus and were sent to a barracks bay. Dinner was an hour or so later, and then it was lights out.

"You'll start in-processing at 9 a.m.," a drill sergeant told us.

I didn't get up when I heard some early morning commotion of people getting ready for the day. Same as the other guys who arrived with me. Ah, a remarkable fifteen more minutes of sleep until the yelling began.

"Get up, privates! You have five minutes to get downstairs for in-processing!"

Promises of extra sleep left my mind quickly. I had moved fast before, but probably never that fast. After a haircut, fresh new duds, and some basic physical training checks, it was on to basic training.

Basic training was no joke and there were plenty of dirt-bags that tried to quit, which put us all through more hell than it was worth to endure the training. I thought about going absent without leave (AWOL), and we had a guy do that and even write us letters saying it was going well. I couldn't do it. I endured the miserably hot Georgia summer, mosquitos, sweat, blood, and occasional tears—for the nine weeks of basic training. After basic training, there were five more weeks of Advanced Infantry Training (AIT), totaling fourteen weeks. I don't remember many of the guys from Basic training except for my battle buddy.

My battle buddy, Tim, was the Army's ideal candidate. He grew up in the country and was a wrestler throughout his school years. He was strong, intelligent, and already had outdoorsman experience. I grew up playing outdoors, but I was also part Nintendo as an '80s and '90s child. I recall asking him for help with simple things like tying knots or dealing with the dreaded Meal, Ready-to-Eat (MREs). I had never had one. I only knew what they looked like. The first one I ever ate was three-cheese tortellini, a vegetarian meal. I didn't know how to heat it, so I ate it cold, which was a huge mistake. MREs taste like someone who just learned how to cook made the meal and messed up the recipe. They're either too bland or too flavorful but in an over-seasoned way. I wasn't eager to eat them again anytime soon.

Tim and I were battle buddies due to the proximity of our last names, so we spent almost every minute of those fourteen weeks together. If it weren't for him, I would have had a slightly more challenging time in basic. Luckily for me, I had him. Luckily for him, I was a quick learner, so I never held him back. We helped each other with everything from making hospital corners on our bed sheets to disassembling and cleaning our M16s before turning them in. At the ranges, we had to run drills together and discuss how we could be faster at whatever shooting lane we were challenged with. During the road marches to and from the range, we would encourage each other by saying, "Just a little further" or "You got it!. Nothing to it but to keep moving!" Chatting with him years later, we concluded that we were all in survival mode and lucky to have one another to help us through that time.

Unlike some movies would suggest, none of the drill sergeants hit us, though they did yell a lot, and sometimes that was funny when directed at someone besides me. We had early mornings physical training (PT) of all sorts and learned weapons, how to shoot, tactics, survival, and many other things, but all at a basic level everyone starts from. A big theme of basic is how to learn to operate as a team, at a bare minimum, and as a unit. This played out immensely in many ways, but one I remember vividly came in the form of a punishment, as we suffered those together, too.

A private decided it would be funny to ask a drill sergeant, "Do you like my haircut?" after one of our routine visits for a head shave. This question didn't go over well at all. After returning to the barracks, we had to *toe the line*. This meant that each of us and our bunkmate had to be standing between the line of bunk beds with your front toes on the line that ran the entire length of the barracks floor at the edge of the bunk bed posts, standing at attention, awaiting further instructions.

Due to the question and the drill sergeants' needing to reestablish that they were not to be talked to like they were buddies, they scuffed us up for over forty-five minutes. I did more push-ups than I ever had, and a pool of sweat formed on the floor in front of me. We also had a "tornado" come through the barracks. A tornado is when the drill sergeants walk through and toss bunks, mattresses, wall lockers, and every bit of clothing into one pile in the center of the floor or throughout the entire bay. We had ten minutes to put everything back to inspection ready. They did drop us a hint, though. They didn't care who had whose stuff; everything had to be inspection-ready by time. We could get our individual items back after the fact. We made the timeline, though I never got all my personal things back.

At the start of basic training, I could barely pass a PT test with a minimum of 60 percent in each of the push-up, sit-up, and two-mile run categories, and by the end of it, I was scoring around 240 out of 300 points. I knew more about weapons, tactics, and fighting than before joining. I was thrilled about my accomplishments but somewhat disappointed as I expected more obstacle courses and hand-to-hand combat training. After all, that's what they always showed in the movies, right? Still, I was proud that I had made it through while improving mentally and physically.

We all survived those fourteen weeks, and some of us went on to our first units. Some went back to their National Guard unit that they signed up through, but Tim and I stayed in Fort Benning to attend US Army Airborne School. The first bastion of awesomeness in our lives and a real rush! Airborne School was also our first freedom from constant oversight; of course, we all used the time wisely to get drunk. We would get as many guys as possible to share a hotel room and party for the night like we were going to war or had just returned from war.

Airborne was a hard three weeks as well. There was daily PT, though evenings were off, and we were free to do whatever we wanted. Throughout the day, we learned the basics of Airborne operations: jumping out of and falling safely from an airplane. We would practice the Parachute Landing Fall (PLF) repeatedly on the ground, then from a platform, and eventually from a swinging trainer.

The trainer was somewhat painful as you would jump off a twelve-foot platform in a mock parachute harness, swing back and forth a few times, and then one of the instructors (referred to as Black Hats, since they wore black hats) would let you drop. It was intended to simulate swinging under an actual parachute and then contacting the ground. If you were lucky, they would release you near the ground. If not, you were released higher and hit harder. Any way you sliced it, you hit like a sack of concrete tossed from a truck bed.

There was also a zipline-style trainer to practice jumping out of the aircraft Again, a mock parachute harness was attached to a zipline, and you would jump out of an aircraft door mockup forty feet in the air to simulate leaving the aircraft It was fun for a few times, but after the fifth time or so, you'd be wondering if you were ever going to find your balls again. Over a few days, we must have jumped out of that thing thirty times.

These precursors of pain all led to the main events. Every day in these delightfully entertaining torture tests, we could see the big tower: a 250-foot-tall tower with four pylons extending out from the top. There were three of these towers on Fort Benning. We saw them every day, whether dropping from a platform or harness. There they were, taunting us, capturing our vision and imagination as we all wondered what it would be like to drop from them in an actual parachute, not just a harnesses attached to torture devices of falling knee, back, arm, and neck pain. The day finally arrived, and they had a demonstration before getting to the first drop. One of the instructors went first while another was safely on the ground with a bullhorn describing what was happening to all the students.

First, the parachute must be rigged into a round metal support that keeps the entire parachute in an open position.

The jumper—or more appropriately at this phase, faller—is attached to the parachute. Then, the whole thing is hoisted to the 250-foot height via cable and pulley system. Once the support reaches the top, it triggers a release for the parachute and begins the descent, giving about ten seconds to figure out which way to slip.

Slipping is pulling one or two of the four parachute support risers as low as you can into your chest to reduce the directional motion of the parachute so that you can fall straight down and perform a PLF upon landing. We watched eagerly as the instructor rode the lift to the top, was released, and began his descent. The instructor on the bullhorn started to inform him which way to slip. The instructor under the parachute pulled the wrong one and began to increase his directional speed instead of reducing it. He hit *so hard* that he was knocked clean out! I looked over at Tim and we both gave each other that look of *yikes*, all while snickering and smiling, knowing our turn was coming.

After ensuring the jumper was OK, the Black Hat on the bullhorn quickly said, "That does *not* happen often, and that is *not* the way to do it!"

After having a few turns dropping from the tower, it still wasn't enough. We wanted to jump from the aircraft You get five total jumps in Airborne School, and I was ecstatic about every single one of them. We learned quickly that everything leading up to a jump sucks— everything! Manifest, pre-jump, getting the parachute, and putting it all on, rucksack, and extra gear if that's a thing for the day. Everyone loved a slick jump where all you wore was the parachute. A combat jump with rucksack, weapons case, and all gear, and especially if you were carrying the radio, and walking with all that gear onto an aircraft, nah man, no one enjoyed that.

Ahh, but when you heard the Jumpmaster start hollering all the usual signals to stand up, hook up, and check equipment, especially after the doors were opened, you could feel that rush of air, knowing that about a minute's worth of sweet relief from all the weight you were carrying was right around the corner. That's what infantry dreams are made of. You knew that you were doing something not everyone in the military gets to do and being paid a little extra to do it!

Shuffling as fast as you could to the door, handing the static line off to the safety, and exiting the door, you had about four seconds of falling into nothing before that static line pulled the parachute out into a fully open position. After checking your chute to ensure it was fully open and functioning correctly, you had about thirty more seconds to look around and enjoy the view of dozens of other jumpers doing and thinking the same thing you were. Then it was time to slip, hit like a brick, er, I mean PLF, and if there was minimal wind, not get drug along the ground while you released one riser of the parachute so it would collapse. It was a feeling like no other.

Tim and I completed Airborne, and then it was on to our next step to becoming like the movie heroes: the Ranger Indoctrination Program—or "RIP," as it was affectionately referred to. This was a two-week assessment to see who could pass and make it into the coveted Ranger Battalions and wear the black beret. We had to wait our turn for the RIP assessment, so while in holdover, we ran . . . a *lot*! Every. Frickin'. Day. You see, when I joined the Army, I weighed 150 pounds. I came out of basic training stronger and weighed 155 pounds. Since I was so light, I could run relatively fast, which was terrific since I was not too fond of running and had never run more than a mile before leaving for basic. We were in holdover for a month, and by this time it was the end of October into November, so it was starting to get a bit cool in the South. I was happy about the change from sweating nonstop.

Everyone has an excuse for why or how they fail at something, lose the big game, or a deal slips through their fingers. The RIP PT test was the first day after a four-day weekend for Thanksgiving. It was my first Thanksgiving away from my family ever. I couldn't afford a plane ticket like most guys could, so I was stuck at Fort Benning for my first actual holiday since joining, turning twenty-one in basic notwithstanding. That weekend, a buddy from basic and I decided we would get a hotel room in Columbus for the weekend and drown our sorrows. Just as we would continue to do after basic training and into Airborne School, it was common for a group of us to share a hotel room, buy a bunch of booze and beer, and drink the weekends away. While we did this under the guise of having a good time, for most of us that Thanksgiving was the first time we'd been away from family during a holiday. I was missing the yearly family tradition of spending the holiday at my grandparents' house. I was alone. We all were, and we decided to cope the best way we knew how: alcohol.

Most dining facilities (DFACs) were closed on post, so we had to get some food on our own. We had Thanksgiving dinner at Waffle House, my first experience with the gloriousness of the Awful Waffle, and engaged in a weekend of beer, garbage food, and laziness.

Not surprisingly, my holidaying left me in no position to score high on a PT test. The RIP PT test also had a higher scoring requirement than the Army minimum of 60 percent in each category. The standard was 80 percent in the eighteen-year-old group. Being a smaller guy, I could run, but ultimately, I failed by one push-up and two sit-ups, or vice-versa. I still tell that story with a snicker of disbelief, as while I was taking the test the grader failed to count four or five of my push-ups and a couple of my sit-ups. Probably rightfully so, as they may have looked sloppy. I don't have a grudge, though. However, I ran the fastest two miles of my life, a twelve-minute forty-second two mile. It didn't matter; what was done was done, and I would not be in a Ranger Battalion. I was down on myself a bit for not spending more time preparing over the four-day weekend.

However, I was still in the Army and would go to an airborne infantry unit regardless. Furthermore, I was more concerned about getting to my first unit and what that life would be like. Most of us expected to go to the 82nd Airborne Infantry at Fort Bragg in North Carolina. What came next was a complete surprise. Shortly after the PT test debacle, fifteen other soldiers and I, who had each failed RIP but scored well enough, were offered the opportunity to go to a long-range surveillance company (LRSC). I had never heard of LRS since I had no military experience before joining and even less knowledge of the military than that.

"Does anyone know what a LRS unit is?" one of the instructors asked. Our blank stares must have given him the answer. "They do HALO (high altitude, low opening) ops, water training with Zodiac boats, rock climbing; they even train for triathlons!"

I was in! It all sounded like bucket list items to check off while getting paid to do it, and to a degree, it was. I accepted the assignment set before me, and so did the other fifteen. We were off to Fort Bragg to be in Foxtrot Company, 51st Infantry, Long Range Surveillance Company, or, F. Co. 51st IN LRSC.

It was mid-December when we arrived at Fort Bragg on a bus we'd ridden for about ten hours from Fort Benning. Some ignorant, joyful optimism had abounded on that ride as we chatted and got to know one another. We had a few days of in-processing before we were released to be picked up by our new unit. Once we were finished with all that administrative garbage, we had a day or two to wait until the unit came to pick us up. Nervous, anxious emotions filled the place over the unexpected. We were lucky to get picked up when we did. Almost all of the company was on Christmas block leave, so it left only a few of the guys around to mess with us.

We received our barracks rooms and roommates, and after that we had a chance to submit our leave forms. I headed back to Washington (state, not DC, which I had never had to specify until joining the military, since most people in the South associate Washington with DC) for Christmas. My emotions ran high as I was thrilled to be headed home for Christmas. Missing my twenty-first birthday and Thanksgiving left me a bit homesick and brokenhearted, but now, after being gone for six months, I would finally be able to enjoy one tradition: a family Christmas.

Christmas leave was great. I saw family and some friends, and I enjoyed hanging out and drinking with the high school buddies I still kept in touch with. Then I returned to North Carolina.

We all arrived back before the New Year four-day weekend because the company commander (CO) wanted the entire unit to come in for a ten-mile run. As the new guys, we were told by the more experienced soldiers, "Don't fall out!" Most anyone who said that looked at us as would a dog with fresh meat in front of them. As the run began, keeping up with the CO was all I wanted to do to prove to everyone else that I could succeed at whatever obstacle was thrown my way. I was somewhere in the mid-back of the pack, and as far as I could gauge, we were around the three-mile mark when people started falling out of the group. I would hear "Close it up" from behind me, referring to the gap in front of me.

I remember smiling a bit with a sense of pride when I saw a few guys who talked smack about us new guys falling out. They had talked down to all of us new guys and couldn't keep up themselves. It was around mile three that the course turned uphill as well. After the long stretch of hill, it flattened out and we slowed a bit to collect the fallouts. Once we were a group again, the CO took off. Guys were falling out again. I'd closed the gap, which was ongoing. Mile five or six was a slowdown to get the fallouts, and then *BAM*, full throttle. This kept going, and I kept up the best I could, but I finally had enough and started to slow back when I saw only about a dozen guys in front of or around me.

I figured it wouldn't hurt if everyone else had slowed back. Besides, I was still near the front. I had no idea where we were, how much longer we had to go, or if this was ever going to end, and that's when I heard the voice behind me: "We're almost home. Don't you dare give up now!" With some verbal motivation to finish strong and a little grit in mental determination, I did just that. Three of the other new guys and I finished with the small group that never fell out. What a welcome to 2001.

As I was increasing in strength, speed, and stamina from only six months of Army life, I felt more confident in my ability to succeed in the physical challenges ahead of me. When we first arrived to the unit, we signed an agreement that stated we would attend and pass Ranger School. It also said that we had two chances to accomplish this goal, or we would volunteer to be reassigned to the 82nd. I looked forward to attending and graduating from Ranger School as my redemption shot for failing out of RIP.

Shortly after the New Year, the water platoon that focused on waterborne operations conducted a two-week, pre-scuba training event to determine which of us new guys would end up in the water platoon. I never thought too highly of myself as a swimmer, but I came to find out I was pretty good. In those two weeks, we did numerous laps in the pool, finning with all the necessary gear: fins, wet suit, face mask, rucksack, and so on, and then moved to open waters at Mott Lake. We performed sub-surface swims in the pool at increasing lengths to fifty meters, drownproofing, treading water with a twenty-pound weight, and underwater knot tying. The fifty-meter sub-surface swim would be the determining factor for who would make the water teams.

I made the fifty meters, barely, without one whisper of breath left in my lungs when I touched that wall. Still, I made it down twenty-five meters with no push off the wall, turned around with the most vigorous push possible, and tried to remain calm while swimming back the second 25. When I knew I had five meters left, I swam as hard and fast as possible until my fingertips touched, and I quickly came out of the water.

I hated treading water with the twenty-pound weight. First, it had to be retrieved from the bottom of the twelve-foot-deep end. Then, I had to swim the weight back to the surface and tread water with only my legs, keeping the weight and my free hand out of the water. Next, I was barraged with questions like, "State your full name, last four of your social security numbers, where you are from," which I had to answer before returning the weight to the bottom of the pool. I could not just drop the weight once this was done, but I was required to swim it back down to the bottom of the pool and come up in a controlled manner.

Drownproofing and knot tying were the easier of the tasks. The others and I had our hands bound behind our backs and feet bound together. In the deep end, we would have to bob in a particular way that was relatively easy to do once we got into a rhythm. Starting with our heads out of the water, we would slowly let all the breath out of our lungs until our feet touched the bottom. Then we'd squat down, jump hard to propel ourselves up high enough that our heads would pop out of the water, take a deep breath, and start the cycle over. It was relaxing once you got used to it.

After ten minutes, it would progress to the dead man float. It's a bit trickier, but it's alright once in rhythm. You would place your body in an L shape with legs dangling down into the water. Your back would be the only thing exposed at this point. When you start to breathe out, you begin to sink, so you'd have to time this one better. The trick was not to let out the air too quickly and sink so that you'd lose your body position but not too slowly that you'd have to blast all the air out and grab the next breath. Remaining calm and getting into a groove was all it took. Tying knots underwater was simply that, tying knots underwater. Either you could do it or you couldn't.

I did well at this entire train-up and was placed on a water team. I wanted to be on a HALO team. Airborne operations were awesome, but we were tethered to a line that pulled the parachute for us after a mere four seconds of falling. I wanted to free-fall for one minute and pull the ripcord for a controllable parachute. However, water ops were a blast too.

It was now sometime around April 2001. North Carolina temps were rising and the greenery was in bloom. My platoon's new task and mission was at the Joint Readiness Training Center (JRTC) in Fort Polk, Louisiana, or, what is commonly referred to as "the armpit of the Army." As the junior guy on the team, I was the assistant radio telephone operator (ARTO). That means I was assistant to the RTO who ran the radios while on mission. I carried many batteries, the backup radio, the antennas, and whatever else the RTO told me to take.

When I went to JRTC, I weighed around 160 pounds. On the first mission we went on, my rucksack weighed 125 pounds, and my kit, the vest we wore to carry ammunition, smoke grenades, grenades—dummies for training; without even the fuses to make a pop sound—and medical items weighed another 35 pounds. I was carrying my body weight in gear.

While I was there, I learned a good deal about the LRS mission. We operate in six-man teams. The team split into two three-man sections. An observation team would go to the surveillance site (SS) near an objective (OBJ), and a hide site (HS) team would remain further away but close enough to support the SS. The SS was comprised of the assistant team leader (ATL), senior scout observer (SSO), and scout observer (SO). The HS team included the team leader (TL), RTO, and ARTO. The scout team would move forward to the SS near the objective, report what they would see to the HS, and make drawings or sector sketches of the OBJ for further use upon return to friendly lines. The HS would report all that information back to the command post (CP) so that the gathered intel could be used to formulate an attack on the enemy objective, whatever it may be that we were observing. It really was that simple.

A mission we conducted was a three-day operation that was completed the same day training ended. End of exercise, or ENDEX, was the magic acronym everyone wanted to hear on the radio. This training mission sucked! It was supposed to be easy: Get a ride on a Huey, get dropped off, walk five clicks (kilometers)—devoid of swamps, creeks, or thickets—to the HS, walk three more clicks for the SS, do two days of surveillance, move to a road and get picked up, go clean all our gear, and go home. Easy. I only had 140 pounds of gear this time.

Not all went to plan, however. The Huey dropped us off at the wrong landing zone (LZ). While that only added a click's worth of walking, it did require cutting through some thick brush and a creek. At least it wasn't a swamp. We got through that and reached the rally point (RP). It took longer than expected to get there, so we decided to stay together as a team and lose a day of surveillance.

LRS teams have one "downed team" mission during each JRTC rotation, and we were the lucky recipients. We were "attacked" with incoming mortar or artillery rounds, which were simulated with the most fantastic fireworks, like oversized M80s. It would be awesome, except it meant that some of us were injured, some critical, and some of our gear was destroyed. It just happened that the destroyed gear was all the long-range radios, and the critically wounded man was the heaviest dude on the team.

The sim rounds went off, and I didn't know what to do.

"Who's injured?" our TL yelled.

The observer controller-trainer (OC-T) started handing out cards that stated our injuries.

"I have a laceration on my forehead," I informed him. "I'm fully functional."

He handed me combat gauze. "Wrap this around your head and help the next guy."

"I have a cut on my finger from shrapnel," another guy stated.

"Great," replied the TL. "Who else has injuries?"

"I have a laceration to my femoral artery!" The heaviest guy on our team, at 200 pounds, had an injury that would require us to carry him.

"Call in a medevac." The TL pointed to me to get the radio up.

"All of your long-range radios have been destroyed," the OC-T chimed in.

"Cross-load the remaining gear: food, the short-range radio, batteries, and other essential items," our TL pointed to me and the RTO, as he and the ATL constructed a litter with branches thick enough to support the weight of our injured man.

Four sturdy tree limbs tied into a rectangular frame and a poncho wrapped around it made for a decent litter. We could wear the four remaining rucksacks, place the extended portions of the limbs across the tops of the rucks, and move out. It was mainly flat terrain for a couple of clicks, which wasn't too bad to carry him in.

We headed west to get closer to the other team on a mission nearby. After six clicks, and we should have been within short range radio of them. During pre-mission planning, we swapped team internal frequencies so we knew if we got close enough to them we could get help or extraction sent to us. Our OC-T had some pity on us, and after about four clicks, when we reached a huge swamp called Big Brushy, he allowed our injured dude to walk while we all carried the litter. I was grateful for the reprieve from carrying him as I didn't know how much longer I would be able to maintain it. We took off at a light jog with him on our shoulders, and I was gasping within minutes. After all the water training, I was in good cardiovascular shape, but the previous weeks of little rest in the field had taken a toll.

We quickly made it through the swampiness, and he was back on our shoulders. We moved him about fifteen clicks since our radio wouldn't reach the other team until we were within two clicks of them. Fortunately, the other team was monitoring their short-range radio. We were ecstatic to be in contact with someone instead of just wandering out in the woods, beyond thrilled to know that this training event was coming to an end, but totally unmotivated when we were informed that we would have to move another four clicks to the extraction point. Long story wrapped up, we loaded up our 200-pound human dirtbag, moved the four more clicks, got extracted via white panel van, and were brought back to the compound. Oh, the 200-pound sandbag that we carried? Yeah, he would have died from his injuries.

We were fresh back from JRTC. Being away for training kept me from drinking heavily, but when we returned I made up for lost time. I got so drunk that I threw up in a boot that was next to my bed, opened the door to the barracks hallway, and tossed the boot down the hall. A guy on my team woke me up the following day and politely informed me that I may want to clean it all up before someone else told me less politely.

I didn't think much of it, and I laughed while doing it. At this point, I had heard all the stories from the older guys, who weren't much older than me anyway. They talked about keg parties on the barracks roofs in their previous units, getting drunk and passing out here or there, and drinking and driving, or motorcycling; all of it only encouraged me and gave me something to shoot for. The stories made me feel like my drinking shenanigans were acceptable and OK. After an incident like this I would usually relax on the drinking a bit, but not for long.

Being on a water team and proving myself in the pool and pond prepared me for the next school. Six of us in the company were selected to attend the Marines Amphibious Reconnaissance School (ARS) at Fort Story, Virginia. The Marines had a RIP, too, but it was a nine-week Recon Indoctrination Program. The course they had to pass to get a spot in their coveted recon community.

The six of us arrived and were ready for whatever this course would throw at us, and it was challenging, to be sure. After meeting with the instructors and fellow Marine students, and getting the gear to the sleeping area that would be our home for the next three weeks, we settled in and rested for the next day. We had some classes on finning techniques, which matched what I had previously learned in our tests to be on a water platoon. Other courses covered our gear, preparing it for waterborne operations, and properly assembling the Zodiac boat for water operations.

The first week consisted of finning in the pool on base at increasing lengths with all the necessary gear. We started at five-hundred meters, and by the end of the week, we worked up to 2000 meters, which is approximately one nautical mile. This would be the new minimum length we would swim for the last two weeks in the ocean. We also learned about charting coordinates using latitude and longitude, which was unfamiliar to me. Until now, I only knew the Military Grid Reference System (MGRS), which used Grid Zones and typically eight-digit grid identifiers to get you within ten meters of a target. Additionally, we learned about tides, charting courses in tides, nautical navigation, and helocasting both ourselves and the Zodiac.

After the first week of classes and pool work, the real fun began. Day one was a nautical mile fin in the ocean with just the finning gear and no extra equipment like weapons or rucks. The instructors charted the course direction based on ebbing or flowing tides. Finning against the current, as we learned, was not enjoyable, as you could watch the shoreline not budge one foot in anything above a one-knot current. This didn't happen until we students started charting the course, and it only occurred twice, thank God.

The other part of this day consisted of what we refer to as motor appreciation. We had to navigate the Zodiac with our six-man team out to a buoy and back using oars, no engine. It's not difficult unless you can't keep paddling in unison and end up zigzagging the whole way, which we did. We were the last team back, but the other teams had eight men per boat. While this made little difference since we used motors after the first day, it made a big difference when we had to carry the boat back to the barracks area. Every. Single. Day. We could never quite get all the water out of the hull, so we just had to endure carrying this thing a little over a half mile down the road, in full uniform, with wet boots, through the sand. Two more guys to help carry would've been gladly welcomed.

Day one was done, and the following week was similar; however, everything had more difficulty except using engines on the boats. The daily fin would add equipment: "rubber duck" rifles, rucksacks that, if not appropriately prepped, would become waterlogged and increase drag, and nautical nav in increasing ranges, all while being rather exhausted from the previous day and feeling we'd not had enough time to recover.

The nautical nav was straightforward. We had to account for the currents traveling from point A to point B. We had to solve for point C. Navigate to where the theoretical point C would be, and the current would naturally take us to point B. In week two, we did all this during the day; week three would introduce us to doing it all at night, yes, including finning, as charted by the students.

During the course, we conducted helocasting. Talk about a free thrill! Jumping out of the back of a moving CH-46 (the Marines' version of Chinook helicopter referred to as a Baby Chinook) at varying speeds and altitudes. We started at ten feet, ten knots. We were given the "go" signal and went out of the back hatch one by one. There was a particular way to jump so that you wouldn't lose your gear or hit the water incorrectly and leave yourself disoriented.

Ten and ten, no problem. We got our team loaded into the Zodiac, back to shore, Humvee (HMMWV) ride to the helicopter landing zone (HLZ), and back at it for twenty and twenty. Twenty feet, twenty knots, what a blast! We were living some theme park fun-time dream and getting paid to do it. Wash, rinse, repeat, and we were ready for thirty and thirty. You know those times when you see accidents at the theme park, and then you think, *"I'm good, I don't need to go on that ride."* That was thirty feet, thirty knots for me. I'm glad they didn't do a forty and forty that day. So, thirty and thirty, ramp open, "GO!" I jumped but allowed my body to get at a slight angle. You want to be straight up and down, and I was tilted about ten degrees. It didn't matter; my feet hit the water and caught, making the rest of my body contact with the water relatively flat. I didn't have the wind knocked out of me, but I was short of breath and had a monster headache for the rest of the day. I'd do it again, but maybe just twenty and twenty.

The final training of helocasting we conducted was tossing a Zodiac out the back, following it into the water, loading up, firing that sucker up, and taking off like we were SEALs on a mission. We did this several times until it was showtime. For the final cast, we tossed that boat out about 300 meters from the Virginia Beach strip, jumped in after it, inflated the rest of the boat, fired up the engines, and skimmed off into the sunset while giving the folks ashore a good show. It's too bad smartphones weren't a thing back then; I'd have liked to see the video.

ARS culminated in a class-wide assault to a point on the beach. It was our last night, and we navigated five or six boats out to a predetermined point and then back to the beach to assault an objective. The instructors had some artillery simulators and some other pyrotechnics going off. At the same time, we assaulted the dunes and lightly wooded areas nearby, which was our objective. It was more akin to fireworks and a congratulatory show of completeness rather than something to react and break contact from. It was a fantastic course, and I was sure I had proven my worth in the company by now with some of the sergeants (SGTs).

The next thing coming down the pipe was training for the Expert Infantryman Badge (EIB). This would give me the clout to prove my worth and lock me in to attend Ranger School. EIB had a two-week train up to the final two days of testing, culminating in a twelve-mile road march with thirty-five pounds of gear plus water. Training started at the end of June, and I was offered two choices: take block leave with the company or train to earn my EIB. I took the latter. We prepared for two weeks on land nav; calling for fire (calling artillery or mortar fire); camouflage techniques; donning gear for nuclear, biological, and chemical (NBC) environments; radios and radio calls; and numerous military weapon systems from the smallest M9 9mm to the M2 50-caliber to the MK-19 automatic grenade launcher. We were up and down these lanes for two weeks, training on all these tasks to be ready to knock it out of the park by test days. Every day was a trial in dehydration and focus in hot and humid temperatures.

Test day came, and the first event was a PT test. Sixty percent or better was required in every category. Only two of the 200+ soldiers that started test day failed the PT test. Next up were the "blue lanes." They had all the tasks split up into red and blue lanes. Red meant weapons; blue meant non-weapons like radios and camo. During testing, you were allowed three total fails for all sixteen to complete. If you failed two events, you were "blade running," which meant one more failure, and you were done.

Grenades earned me my first failure. You had five grenades to hit three targets. My first one was a thirty-five-meter throw, and it had to be within a five-meter white circle to count. I threw the grenade short, planning for it to roll in. It didn't. It hit and stopped dead. The second throw was in the circle; three grenades remained. The second target took me two attempts, and finally, the bunker. I had to prep the grenade from the side of the bunker while in the prone position (on my belly), without any part of my body forward of the edge of the bunker, "flagging" myself. Once the grenade was prepped, I had to lean forward slightly to peek at my target, the opening of the bunker where the machine gun would point outward. Then, I had to reach and drop my grenade into the bunker. I did all that and heard a *thunk* as my grenade bounced off the front of the bunker and rolled away on the ground in front of me. Fear not; it was a dummy grenade, but knowing I failed the task was akin to seeing a live grenade roll past anyway. Mentally pushing my failure on my first event aside, I quickly reassessed what I needed to do the next round and got back in line for a retest. I got my retest and passed the second try. Still, I would have rather had that retest in my pocket for later.

Later that day, I jacked up the call for fire. I tried to get a freebie redo by saying the instructor gave me the wrong directions. But I messed up by claiming that he messed up, and it only made it look like I messed up, so the whole thing was messed up. This test became my second no-go.

Day two, and I was now blade running from having failed two tasks the first time. I passed all the weapon systems just fine. I completed all the tasks while blade-running and only had the twelve-miler to go.

A twelve-mile road march didn't sound too bad. But then there was the thirty-five-pound rucksack, before adding water. The uniform and garbage military boots didn't help the situation. Nor did the weight of the helmet. We started at 1:00 a.m., and even though it was early in the morning it was still 70 degrees with 75 percent humidity. It was July 19, my twenty-second birthday. I had spent my twenty-first birthday in basic training, and now, my twenty-second was doing EIB. It had rained the night before, so all the sandy trails it followed along were soft, and some areas were water-logged. There was only about one mile in total along asphalt. Finally, it had to be completed in three hours. It was the most brutal twelve-miler I ever completed for time.

Of the 200 dudes that started testing, about 120 made it to the road march. Of the twenty guys in our company that started, only twelve were left to do the march. We were lined up and ready to go. An official came through to verify that everyone had the required load and that our rucks weighed at least thirty-five pounds. Once complete, we were sent on our way, and the timer started when the last guy got through the gate. I saw some guys dump water as soon as they got past the starting point to reduce weight, but I didn't do that. I knew I might need it later.

I picked up a good jog after I went through the start gate and kept that for a while. I had yet to count on the environmental conditions to get to me so quickly. By mile two, I was already drenched in sweat. My endurance was holding up just fine. Mile four started at the bottom of some low ground that was washed out by water and made almost a quicksand-like mixture that was challenging to get through, followed by a rather steep uphill climb. What's worse is that it would return around mile eight. I got to the asphalt portion at the top of the hill, but my knees were already a bit beat, so I stayed to the side in a worn-down dirt path and jogged as fast as possible.

Mile six, the turnaround point. I freaked out a bit as I saw the timer show 1:20. I was at the halfway point, and it took me an hour and forty minutes to get there. The timer wasn't counting down as I had thought; I had only been marching for an hour and twenty minutes, but I was greatly concerned, thinking I was over halfway on time at the six-mile mark. Panicked, I picked up my pace and grabbed a cup of water being handed to me. I had plenty of water, so I took the cup of what I thought was water and splashed it on my face while running. It wasn't water; it was orange juice. Combined with the sweat pouring from my head, it ran into my eyes, burning and stinging them. I was enraged.

I yelled a few curse words and was fortunate enough to pass another table with cups of water. I splashed that on my face, wiped away what OJ I could, and kept apace. At mile eight, I was exhausted, dehydrated, and ready for this thing to end. The top of the hill bottomed out in the quicksand pit. I used the steepness to my advantage and ran as fast as possible to the bottom. Somehow, the watery sand felt worse this time; it was probably my body more than anything. Everyone walking through it had churned it into a softer mess than the first time. My hamstrings and quads felt tight, and my calves cramped as I walked through it.

Is this how I'm going out, cramps? I thought as I tried to push through the pain and not focus on it, but it gripped my mind and had my full attention. I decided that if I couldn't finish this road march, it would be because my body quit on me and not me stopping for the pain. I picked my pace back up, and slowly, the cramps dissipated. Around this time, instructors were marking the miles. They didn't tell us how much time we had left, only that we would make it in time if we kept pace.

In the end, I finished with eight minutes to spare. My brain and body were fried when I got to the table to check in and be counted as completed.

"What unit?" some sergeant asked me. Whatever he said didn't register in my brain.

"What?" I replied.

He turned me to the side, looked at my unit patch, and said, "Oh, one of the LRS guys, right? Four of your guys have finished already. Join them."

I found my group, and out of us twelve, eight completed the road march. I had earned my EIB, but was I finally in the club? Hell no! I still had a seat at the kids' table. At least it was at the head of the kids' table.

Happy birthday to me. I got my EIB stabbed into my chest, as infantrymen like to do to one another.

Two days after the road march, my feet still hurt. My buddy Derreck was hanging out in my room playing video games, and I sat on a chair rubbing my sore feet and drinking way too much to celebrate my EIB. Eventually, I got up, told him he could stay and play as long as he wanted, and went to bed. Within ten minutes, I was on my back, vomiting on myself. He came over and rolled me to my side to make sure I didn't choke on or suffocate on any of it. He and my roommate Bryan took me to the shower, turned it on, placed me in there, and helped me wash up. They said that I said, in a joyful and kiddish manner afterward, "I'm clean!" They took all the blankets off my bed, put them in the hallway, and watched me for another half-hour to ensure I didn't do it again. I didn't remember any of it the following morning. I had bought a fifth of Jack Daniel's to share with whoever came by to have a shot with me. Apparently, I drank almost the whole bottle by myself.

After EIB, I was ready to take leave, and after being given a three-day weekend for my success, I returned to work with an offer. I could submit my leave form and take leave, which I was very much ready for, or, I could begin training to be an emergency medical technician basic (EMT-B). Since we operated in six-man teams, it was recommended for us to have two EMT-qualified personnel per team. This meant we could attend the course, so I took the opportunity. I had eight weeks of nothing but medical training and was left to my own devices. I couldn't pass it up.

This was the first course I'd done in the military that required me to be studious, and I did rather well in EMT training. From the book knowledge to the hands-on portion of analyzing a casualty or even just a person in pain, I did no better or worse than the medics in the course with us. They had prior experience and knowledge to fall back on. Most of them had no problem helping us in any area we lacked. We got to shadow in the ER at Womack Army Medical Center and even go on ride-alongs in ambulances with EMTs. I became a nationally registered EMT-B with Basic Trauma Life Support (BTLS) certification.

There isn't much else for me to add, but if you've been paying attention to the timeline, you realize that something else happened around this time: September 11, 2001–9/11.

During the last week of class, one of the instructors stopped us mid-instruction and turned on the TV mounted on the wall in the classroom. The first tower had already been hit, and the news was on it. Aside from the TV, you could hear a pin drop. We were glued to the TV in mixed thoughts, emotions, and disbelief. I'm sure this conjures up those feelings as you read. A little chatter here and there, "Think they'll send us?" "How did this happen?" "Who did this and why?" And on and on. Then *BAM*! The second plane hits in real-time, right in front of us. Back to jaws agape and hearing pins drop. Time stopped. I didn't know what to think or feel. Stunned silence.

Then, the towers collapsed, one at a time. How could this happen? I didn't know anyone in New York but that didn't change the fact that this wasn't supposed to happen. I knew it meant we would be at war, but not with whom or for how long? *Would we be deploying to some foreign country in the coming months? Was all my prior training enough?* So many questions were running through everyone's minds as we stood in silence in grief-stricken awe.

I don't know how long we all watched until the lead instructor finally, reluctantly, turned the volume down and turned to address the class. He asked if anyone had family in New York, and if they'd like to be excused to attempt to contact them, but no one got up. He said that while it was tragic, we still had instructions to get through, but they would try to get through it as quickly as possible. He kept to his word, and we finished the day early by lunchtime.

The next day, we drove over to class, and it took two and a half hours to drive the four miles to get to where class was held. Fort Bragg wasn't taking any chances, and even with ID cards, they searched every vehicle. Get out, pop the hood and trunk, and look in consoles, glove boxes, and mirrors under the car—that type of search. Understandably so. Army posts used to be drive-throughs without security gates, guards, or anything. That changed forever. The following days, since we only had three more days until the training was complete, we decided to ruck march over there for our PT and change at the gym near our classroom. We wrapped up EMT training and waited our turn to go to war.

I still couldn't take leave since we were preparing to go to war. I kept my eye on the prize, that Ranger tab. By this point, I had proven my worth, and I had already done a lot more in one year in the Army than some do in an entire career.

I was sent to the Pre-Ranger Course (PRC) a month later. Fort Bragg had a two-weeklong PRC which was supposed to prep individuals for the rigors of Ranger School. There was challenging PT, reduced amounts of sleep, and planning classes for squad and platoon-sized elements to conduct patrols, recons, ambushes, and raids. It culminated in a three-day field exercise in squad-level tactics. Multiple daily recitations of the Ranger Creed were a must, with punishments for every mistake. There was not a single day that we made it through the entirety of the Creed without an error and numerous push-ups or flutter kicks. PT usually involved carrying multiple forty-five-pound water cans with us everywhere we went.

It was November, and the nights were cold, but the days were fine. Midway through the field exercise, one of the instructors allowed us to start a fire as we gathered around and talked about the day prior, the rights and wrongs, dos and don'ts, and what to get right for the second half of the exercise. The Army has its After-Action Review (AAR) for everything. After we got through the formality, the instructor asked what we wanted to know about what was happening in the outside world. Though it was only ten days, plenty had gone on.

"What is going on with the Middle East?" We asked.

"It must still be in the planning phases since we haven't heard much yet," he informed us.

PRC was a blip of a thing to complete before it was showtime: Ranger School back in Fort Benning. My time was coming, and I knew I was as ready as ever.

Thanksgiving came and went. I stayed with some buddies in Athens, Georgia, since I couldn't get a ticket home. Due to the deployment uncertainty, mileage passes and leaves were restricted to 500 miles. Again, I drank and ate my way through the holiday to compensate for a probable underlying depression about being away from family. I felt terrible since I drank so much on Thanksgiving Day that I threw up on the floor by the bed of the spare room I was staying in. My buddy cleaned it up while I slept off a hangover the following day. I apologized profusely to him as I was embarrassed that I was so drunk that I threw up at his parents' house.

"I was up chatting with my mom, and we heard you. I thought you were in the bathroom, not beside the bed," he said.

We did go to some bars, and as in previous times, drinking always seemed to be encouraged, at least acceptable, but I had never known when to say when.

It was January 2002, and I was supposed to go to Ranger School in one week. I tried to work out and run while on leave, but it was Christmas, and it didn't happen. I did *not* learn my lesson from RIP. Well, I did, but it was challenging to make PT a priority. Fate would have it that something was wrong with my paperwork, and I wouldn't start Ranger School until February. PT was on, and I used the extra time to prepare. Then I threw my back out.

I'd had some back-and-forth issues with my lower back already, but this time was worse. I couldn't run or do a sit-up. Push-ups hurt, and I couldn't even put my boot on without trying to hold it by the tongue and fling it over the end of my foot. The diagnosis was piriformis syndrome. A muscle connects the hip bone to the greater trochanter, the top of the femur, which controls the rotation of your foot inward and outward. This muscle was staying contracted, and it was constantly pinching my sciatic nerve and causing all sorts of pain. The Army dealt with that the way the world did at the time, Percocet and stretching. It put a significant damper on my training, but the stretching helped, and by February I was doing well enough to attend. I could at least get my boots on with minimal pain. I was worried that one wrong move would send it into another spasm, and my Ranger School debut would be short-lived. Plus, the platoon leadership at the time would *not* let me wait one more month. I was going, end of story.

Ranger School includes a "Hell Week" of grueling physical events to weed out the weak. The PT test was first, with 80 percent required achievement in each category. Here was my opportunity for redemption. I passed the push-ups test just fine, even with the instructor not counting some.

One event down, two to go. I failed the sit-ups on my first try by something big like six, and he counted every single one! We were allowed ten minutes to rest before the re-test; it was more like five, and I didn't think I would make it. One of the Ranger instructors (RIs) said, "You're all warmed up now. Go knock it out!" Somehow, that gave me some confidence, and I knocked it out with about fifteen seconds to spare. I should have done that the first time! In my younger years, I was constantly maxing out the run, which was thirteen minutes or better for two miles. I hit thirteen minutes on the nose. PT test done.

Within that week, we had to finish a five-mile run in less than forty minutes, which was instructor-led. While I didn't have a watch, I know we did it in less than forty minutes. We had to complete a 2.5-mile equipment run in under twenty-four minutes. There were two obstacle courses and a water confidence course in the Benning Phase. Malvesti was a short, half-mile obstacle course that was relatively easy to complete. The other obstacle course was the mile-long Darby Queen. The water confidence obstacle was a huge challenge to those fearing heights and requiring great balance to not fall off and into the icy water below.

During all these physical events, we were yelled at routinely for being too slow at something or other, taught patrolling, recon, and ambush classes, deprived of sleep, hurried through the DFAC like at basic training, and expected to retain all this newly acquired info to use in the field.

While in Ranger School, every challenge we faced be-came more complex with sleep and food deprivation, more weight on our backs, or more personnel to be responsible for. I wanted to quit numerous times due to my failure or, more likely, someone else failing me. Throughout all the physical and mental exhaustion of all the trials thrown our way, we all displayed the intestinal fortitude to fight on to the Ranger objective and graduate, which came with much elation, relief, and some concern knowing that we were at the top of the rung now. The junior enlisted would be looking at us to lead them into the same success we had just accomplished.

With Ranger School completed, all gear turned in, and nothing to do but to sit back and enjoy our success with food that didn't come from a brown plastic bag, it was time for The Gator Lounge and a couple of beers, but not for me. I spent the next two days in the Troop Medical Clinic (TMC) due to cellulitis in my knee. I had a small cut that had gotten infected in swamp water and needed antibiotics. It could have been worse, though; some people with wounds like mine need the area cut open and scraped out, then packed with gauze, which I hear is painful. I only needed an IV and antibiotics, but I had to stay in the TMC. No Gator Lounge, no beer. However, some guys brought me good food and ice cream, which was a win.

When we headed back to Benning for graduation, another guy and I had to ride in a car with a medical person present, so as soon as we got in the car, he informed us the trip would take about six hours and we could sleep if we wanted. I fell asleep in about three and a half minutes, the same for the other guy with me. We stopped at a gas station, and I ate two Burger King burgers in about three minutes. Then, we were back in the car and out for the rest of the trip.

Back at Benning, it was chill until graduation. Ranger School graduation is something to see for sure. Look up "Rangers in Action," and you'll see what I mean. They conduct a live action recreation of everything we did in about a sixty-minute show for all the families to see and make it all look sexy while doing it. It was cool, but I wanted to hit the road. Derreck recycled from the class ahead of me, and we would ride back together in my car; we were ready to go.

The show wrapped up with us receiving that gold and yellow tab. Two non-commissioned officers (NCOs) from our company made the trip down to pin on our tabs. What a feeling of pride, accomplishment, and freedom. We stopped for lunch with the two NCOs and then returned to Bragg. We stopped roughly every hour to get energy drinks and candy bars. I started Ranger School at 165 pounds and by graduation I was back at my weight upon joining the Army, 150 pounds. The funny part is it only took one week to get back to 165 pounds. I was glad it was over, but it was still just the beginning. We had proven our worth and had a seat at the table. More opportunities opened up, and I wasn't even two years in yet! It was April 19, 2002, and I had earned the coveted Ranger tab and, in my eyes, redeemed my previous RIP failure.

My sleep schedule was out of whack for a good while. I went to sleep that first night but woke up around 3:00 a.m. I tossed and turned for a bit, trying to get back to sleep, but to no avail. I decided to go upstairs to see if Derreck might have the same issue. As I approached his door and planned to knock quietly, I could hear a sound like the clicking of a video game controller. I listened for a few seconds to ensure I was hearing it correctly, and when I was sure that's what I was hearing, I knocked. I heard, "Yo," and opened the door.

"Can't sleep either?" I asked, snickering.

"Hell, no!"

"Want to go to Waffle House?" I grinned.

"Hell, yeah!"

We had the small table covered with food and annihilated every last bit. The following night was the same, but I woke up closer to 5:00 a.m. I went upstairs and heard Derreck playing video games again, so I knocked and got the same reply.

"What time does Cracker Barrel open?"

"Dude, I used to work at one in Texas. They open at 6:00."

"Let's go."

We ordered so much food the waitress said, "You boys ain't gonna be able to put away all that food!" We smiled and cleared every plate we had, much to her surprise. That's how I gained the weight back. We weren't starving anymore.

It was unbelievable how completing Ranger School and returning with a Ranger tab instantly changed the attitude of others toward me. As I mentioned, I finally had a seat at the table, and doors were about to open. There was a power trip that I went through. I was also a lonely person. As Derreck and I were the first to return with tabs, we were even more of a hot commodity than our non-tabbed compatriots. We had proven we had what it takes to endure hardships with leadership qualities that would surpass any difficulties thrown our way.

All my friends didn't have tabs, which meant that in the work environment, they were beneath me. I had to "smoke" (haze) my buddies at work. Smoking someone was a day-to-day reality; I hated it, so I didn't want to do it much either. I would usually apologize after work with alcohol since I was over twenty-one and most of them were underage. Everyone understood that if they were in my shoes, they would be expected to do the same thing. That didn't make me feel any better about it as I really hated doing that to my friends.

Am I losing myself? I wondered, as my drinking picked up even more.

CHAPTER THREE

The company promoted me to corporal, but all this meant was that I couldn't hang out with my buddies. Now I outranked them, and as a result, would have to give them orders. Creating a divide between my buddies and me at work would help establish that they had to take orders from me instead of ignoring what I told them to do. The NCOs didn't want me around since I wasn't a true NCO; I wasn't a sergeant yet. Sergeants had more responsibility as it was the first rank where they were held accountable for leading people, more than just themselves.

They also started a draft in the company and fought to split us up into other teams. I went to another team that traded me for a lesser soldier. I didn't want to go to another team but my frustration about it wouldn't last long as I was about to cross off a bucket list dream. I ended up on a HALO team and was ecstatic for the opportunity. I liked the water training, but who wouldn't want to free fall *and* get paid a little extra for it? Airborne meant an additional $150 per month, but HALO meant $225. That was big money to a corporal twenty-plus years ago.

The people on the team where I moved were a bunch of assholes. No, seriously, they were possibly the biggest assholes in the company. All were from other infantry units, one from a Ranger Battalion, and all had chips on their shoulders, which I would soon develop. I was scared going about on to this team, as I knew about them and avoided them when possible. Walking in as the new guy, I expected to get smoked, choked out, berated, or something similar. Nothing like that ever happened. They welcomed me like I was a long-lost brother. They told me what they expected of me, and in turn, I was rewarded with corporal (CPL) rank and HALO school.

All I had to do was be the pit bull. If someone messed up, it went through me first. If I took care of it, "Good job, Chris!" If I missed it and it went to an NCO, "Why did you mess that up, Chris?" That's when I turned into a colossal asshole myself. I didn't want negative attention on me, so I would get on everyone else. At least it looked like I was doing what they wanted. If someone didn't have a required uniform item, I'd have them doing push-ups, flutter kicks, or whatever other punishment I could think of until they began to sweat. A couple minutes late? I owned you. I hated doing it, but not as much as I hated feeling like a failure when the NCOs would get on me about someone else's failure.

I drank more because I didn't like who I was becoming. I was drinking at least a six-pack per night—if not eight beers—and then a case or more on the weekend. I was a functioning drunk in the evenings, pretending I was OK, and the drunkenness caused me to be an asshole outside of work, though unintentionally. In my drunkenness, I would end up in someone's room, messing with them or their stuff, and they would tolerate me and try to get me to leave. It wasn't every night, but it was most nights. No one punched me in the face, but I tell you, I deserved it, and it would've been a wake-up call.

I'm sure most of these guys were relieved to know I would be out of their hair for a while as I started HALO school. The first week was at Bragg. We learned how to fly our body in free fall with hours in the wind tunnel and how to pack a parachute. The wind tunnel is exhilarating fun, for sure. After leaning in with your body in the proper free fall position and getting tossed out numerous times, you develop a feel for how the wind flows around your body. You learn how to tilt your torso and arms to make turns in the air. Each arm movement requires a countermovement from the other arm, so when you reach in to pull the ripcord, your opposite arm has to move up over your head like a Taekwondo high block; if not, you run the risk of rolling over as you deploy your chute.

Once we were comfortable with this, they added a rucksack. You can mount a rucksack front or rear. The rear mount was easier to control, but there was the risk of the deployed chute getting wrapped up in the gear. The front mount could shift and cause spins or rolls that made it hard to control, but the gear was in front and out of the way of deploying chutes. We practiced both, and then there was oxygen equipment too. Jumping above 18,000 feet required supplemental oxygen, so we practiced while wearing a mask and oxygen bottle strapped to the side. After the week of this and being proficient in packing a chute in twenty minutes or less, it was on to Yuma, Arizona, for three weeks of jumping.

In the second week of June 2002, it was 115 degrees in Yuma daily. We caught a bus at 3:30 a.m. to go to a terminal, listen to pre-jump, practice the emergency procedures in a hanging harness rig, and get ready for the jumps ahead. There were specific gates to pass. For example, by the fourth jump, you had to have a stable exit, perform a proper left turn and right turn, and count six, five, and four. That is to say, at 6,000 feet, you had to clear the airspace above and below you by looking around; at 5,000 feet, you had to wave over your head to signify you were going to deploy your chute and then deploy the chute at 4,000 feet. Jump three was a practice test; if you passed, jump four was a fun jump for extra free fall time. The wind tunnel helped us learn how to free fall, but the exit was on you to figure out with recommendations from the instructor. In those first jumps, you jumped one-on-one with an instructor who would coach you through all these steps if you had any issues with them. You arch your back when exiting and in free fall, but that's not a good description. You push your pelvis forward to become a cone shape, forcing the air to bleed off you in all directions. Think of a badminton birdie. If you sit while in the air, you'll flip over, and your ass will become the area the air will bleed off from. I could fall just fine, but I couldn't nail the exit.

On the first jump, I got to the edge of the ramp of the Sherpa aircraft (C-31, I believe) that we used most of the time. There, I prepared to hurl my body into the great sea of blue and the tan colored sand. Looking at the horizon, I kept my head up and jumped and then . . . I don't know what happened. I tumbled and arched, and then I was in free fall with the instructor giving me hand and arm signals. I did the corresponding action to the signals as I remember seeing them, and then it was six, five, four, and I was under the parachute. The drop zone (DZ) was right in the area, so I tried to follow those already under me. We all landed near but off the actual DZ. I was relieved to live through the ordeal and was eager and anxious to hear what the instructor would tell me. All gear in hand, we loaded a bus back to the terminal, repacked our chutes, got our debrief, and got ready for the next jump. It's a crazy thing to have to wake up early and deal with altitude changes, then an enormous adrenaline rush and crash, and the heat and sweat, and to pack a chute. It's extraneous when you haven't lived that lifestyle and then have to wait to do it again multiple times per day. The body goes through so many changes quickly; it's better than any high, including flying a helicopter. As

I mentioned, my exit was nonexistent, something akin to throwing myself out and hoping for the best. Other than that, my free fall was on par for new students.

Jumps two and three were similar: a tumbling exit and decent free fall. If you failed jump four, you got a retest or two before subjection to a commander decision, which usually meant you'd be sent home. I did *not* want that. My instructor suggested walking up to the edge, keeping my eyes on the horizon, raising like performing a calf raise, lowering back down, and then jumping while counting one, two, and three. One, calf raise; two, back down to flat foot; three, jump. That night, I was so anxious and concerned that I would be headed home that I decided to practice diving onto the bed in my hotel room until I was confident I would nail the exit. I stood at the foot of the bed, did the count he recommended, and jumped with my pelvis thrust as far forward as I could so that it hit the bed first before any other body part. They referred to the position as "dick to the door," so that's what I did. I jumped on that bed so many times my belly had to be red. I went to bed exhausted but more confident that I would get the exit tomorrow.

Jump four: put up, or shut up. Before the bus came, I practiced a few more times and tried to envision the ramp with the horizon. I arrived at the terminal, pre-jump, emergency procedures, ready for first daylight and to knock this jump out. It may have helped my cause, but for some reason, that day they had a C-17 that we were jumping from. If you don't know, that's an enormous aircraft It will hold a ton of cargo or a ton of people. It's so big that it creates this dead air space around the plane that prevents you from feeling like the hawked loogie you are for about a solid second, a remarkable phenomenon. I liked every C-17 jump I ever had over the years.

Anyway, I walked up to the edge with my instructor, did the count I had practiced, and jumped. I nailed that exit and every one after that. The instructor was also overjoyed. He was happy, as it meant he was a good instructor who had helped me through my deficiencies. I only needed to do that count another time or two before I'd be jumping with more equipment and simultaneously with more people. After jump four, you start jumping with two students to one instructor, and fewer instructors are required the more you progress. We moved into more equipment, oxygen, group jumps, and formations under night conditions. I liked nights since I didn't have to wake up at 3:30 a.m. anymore, and going into the night with the temperature dropping was better than going into the day with the temperature rising. If I could pinpoint one moment in my life that was still, to this day, the most fun I have ever had in the Army, it would be that month in June to July 2002. HALO school was complete with twenty-five jumps. It was on to leave, and then my second trip to the dreaded JRTC.

August 2002: JRTC. I went from the most extraordinary experience in dry heat to the most miserable experience in humid nastiness that I can fathom. One day on the radio, we heard, "Today will be 100 degrees with 100 percent humidity, so make sure you're drinking that water." *DAMMIT!* Ticks, chiggers, heat exhaustion, and prickly heat. Have you ever had that? It's when your pores become so clogged from all the crap in a place combined with sweat that they crystallize shut, and when you sweat, it can't escape, so you feel itchy all over your back or wherever it is on your body. Here's how I learned about that on the one mission we had.

It was supposed to be a five-day mission. Provide surveillance on an objective three days before a battalion, or whatever size element, strikes an enemy village. This mission went to shit before we even got off the helicopter.

They still used Hueys at JRTC, and they are some excellent aircraft, though very dated. They have a distinctive *wop-wop-wop* sound from the two-blade rotor system and are just iconic in the world of rotary wing aviation. As we approached the HLZ, our TL tossed his ruck off his lap, and we followed his lead. It's a good thing we didn't jump immediately following as he thought it was ten feet, but it was closer to forty feet! After landing and un-assing the bird, we found our gear and all the equipment that had fallen from the drop.

Luckily, nothing was broken. We got on azimuth and began to make our way to the HS. Right off the HLZ, we encountered some thick brush that consisted of "wait a minute" vines. The vines' name refers to how they hold you up when you're trying to get through them. In the thick of this, one of our guys lost his boonie hat with the multiple integrated laser engagement system gear attached. MILES gear was used to determine whether somebody shot you and was a sensitive item that, if lost, there was hell to pay. We had to use a white light to find it, which violates all tactical sense, but we were in the thick of it all and needed the extra light to see it. We finally arrived at the HS and sent the SS team to do their thing. As a corporal, and with our sergeant stuck in the rear suffering from an allergic reaction, I was in charge of the HS. I was good with the radios and decent with HS construction, so it wasn't a big deal. We got lucky, and the enemy was behind schedule; otherwise, we would've been captured at some point, and I would've had more to answer for.

It was day two and we were spent from the BS movement and all that crap trying to get there. I didn't know what zero illum was at the time, but that's what we had, which means the moon was down and not providing any nocturnal illumination. We couldn't see to dig a hide site, but we tried.

It was so dark we tried digging with night vision goggles (NVGs) on and still couldn't see a damn thing. We got something of a hole dug, and I put two of our guys in it, covered it with whatever I could find, and pulled watch the rest of the night. The OC-T with us was not impressed, and he let it be known, albeit relatively gently. They don't have authority out there, but he was trying to encourage me to do more, so I tried to find the best place to camo us into some low ground. Ultimately, it worked, but I'm glad the SS reported no enemy. In hide and seek, we may have done OK. We wouldn't have fared very well with opposing forces (OP-FOR), who would be motivated by rewards for finding us. It rained off and on the entire time, and then the battlefield moved so fast that they pulled us a day early.

We were all thrilled to be moving to extraction, but exfil proved to be just as much of a mess as every damn day had been up to this point. We linked up and began the movement, five or six clicks to freedom from this debacle of a mission. The terrain was roughly straight and level for the entirety of the movement. Being near main avenues of approach, or hard surface roads for the civilian puke, caused us to redirect multiple times. Oh, and of course, it was raining. It was mainly a light rain, but there were some moments of straight downpours. At this point, none of us had been dry at any point during the op. The illum was still low, so NVGs weren't any help either. We all had a turn at a stumble and fall, though we were all rather proficient in walking at night under these conditions. What made matters worse for me at one point was the rain was knocking something out of the trees, and I had allergies flaring up most of the time. This manifested at one point when I didn't see a tree branch, the pine needles brushed against my face and eyes, and while nothing poked my eye, something came off and was in both my eyes. My eyes swelled shut, and I couldn't see for about five minutes. Our TL didn't want to stop and told me to walk it off. *Yeah dickhead, walk it off when I can't see!* I kept blinking as best I could and catching handfuls of rain to splash it in my eyes, which helped, and eventually, I was back to good. I was freaking out, though, thinking I was going blind or something and wondering how long it would last.

Slips, trips, falls, rain, and temporary blindness aside, we finally reached the extraction point. We were getting picked up by a Black Hawk this time, and all I could think about was looking forward to sitting by the open door and letting the wind beat the hell out of me until I was dry. I liked the "hurricane seat," which received all the down-wash from the rotor, as I enjoyed the wind in my face. One more obstacle before extraction; that effing prickly heat I mentioned a few paragraphs ago. It was so bad on all of our backs that it was driving us into a lack of tactical awareness or even caring at this point. One of our guys just stood up, took his battle dress uniform (BDU) top and t-shirt off, and started to scratch his back with them, like how you'd dry it with a towel. Mind you, he was standing right by a road with no care for the potential of enemy presence or compromise. The TL yelled at him as we were still "in play" and then yelled at me, being admin ATL, for not fixing the situation myself. The truth was, I wanted to do the same thing.

The Black Hawk came and transported us back to the CP. Once we debriefed, probably looking and smelling like wet dogs, we went and grabbed some buckets, headed for the water buffalo, and proceeded to strip and wash right there. Having someone scrubbing your back in the dark of the woods is a weird feeling to help clean your super-clogged pores so you can feel better and then return the favor. Someone had some witch hazel, which helps to open clogged pores. After cleaning up, we all hit a cot to dry our feet. They looked like eighty-year-old feet after a three-hour bath, I'm assuming. Our CO walked in, saw our sad state, and let us know that would be our only mission for the duration of that clown show. The only good news we had, I'd say.

JRTC was finally complete. We were back at Bragg, and we had a new platoon sergeant (PSG). I was a hothead and trying to prove myself. I was already tired of being a corporal, as all the sergeants and above would try to ditch their duties on me, which needed an NCO. I wanted stripes and would do anything to rub it in their smug faces that I was better than them. I was at the lowest point in my life because everything I did was out of arrogance and pride. These qualities can make you successful in infantry-land but are qualities that, if left unchecked, lead to conduct unbecoming of a decent man.

I decided my subsequent pursuit would be Jumpmaster School (JM). Passing JM at Bragg means you're the cream of the JM crop. If you pass JM at another base and then come to Bragg, you must take Bragg's JM refresher, as they don't hold the other JM Schools in high esteem. It isn't the same the other way around. That's how detailed and respected Bragg's JM School is. However, you have to pass Pre-JM first.

I didn't respect all these NCOs anyway, and they didn't want to waste the three weeks with the demands of JM School. The easiest way out was to fail Pre-JM, and then you would have to wait two or three months before you could do Pre-JM again. I saw an opportunity to stick it to them, so I took it. Backing up a bit, when we got our new PSG, he sat down with us and asked about our goals. I was cocky AF, and I told him I wanted stripes. He informed me that I could attend JM as a CPL and said he would get me on the promotion board if I passed JM. I was so cocky I said, "Done!" and began prepping for Pre-JM. All you had to do to pass Pre-JM was rig a rucksack and an M1950 weapons case in jump configuration within fifteen minutes. I may have been a drunk, but I was a tenacious drunk. I practiced at work, in the barracks, when I was sober or intoxicated, and whenever I had a chance I would rig those two items up so that I had no doubt I was ready. Then it was test day.

Hundreds of soldiers were there for Pre-JM, including ten NCOs from our company. I hated most of those guys because they had years of experience over me and still acted like shitbags. I couldn't stand them and I felt they were subpar NCOs, but due to their position, rank, and prior experience, I could only talk so much shit until they treated me like a private for acting up. I had just over two years in and was ready for advancement; here was my chance.

Once my group was in and the timer started, I began rigging the rucksack first, as that took longer, and then I started the weapons case. We weren't allowed to have watches on, and smartphones weren't a thing, so I had no sense of time. I got done, looked around, and saw many people still rigging. I thought I was too quick, so I double-checked my rigging. I took some time to double-check everything; from what I could see, I was good to go.

People were still rigging, and I got nervous that I'd messed something up or missed something entirely. Did I do this too many times drunk? I triple-checked, and shortly after my quick final check, I figured, eff it; if I missed it, I'll do it again another time.

"Time's up!" Here comes the Black Hat (like in Airborne School, there were Black Hats in JM) to inspect my ruck and M1950. He looked them all over, said I was a "Go," and handed me my pass slip. I headed back to the company with a huge smile. I saw the other NCOs, and none, not *one* of them mother fuckers had passed!

As I returned to the company, my heart raced; I was out of breath and a little heated. I was so excited to see the PSG and let him know everything. We were having a

Battalion Organizational Day (BN ORG Day: "mandatory fun;" families invited), so everyone was in civies. However, work still had to be done, and some of the senior NCOs were still tending to admin crap. I ran up the stairs, into the building and his office, and showed him my slip. He smiled and congratulated me as I told him that wasn't the best part. I informed him that I was the only one in the company that passed and that I knew a few planned to fail so they wouldn't have to endure the rigors of JM. His smile turned flat and then to a disgusted look. He jumped out of his chair, went to the common area where all the PSG offices were, and announced loudly how his CPL was the *only* NCO in the company to pass Pre-JM! He did some shit talking about how the rest of these "Legs" (derogatory slang for non-airborne personnel) should have their jump pay revoked for not passing Pre-JM or having the desire to become a JM, especially at a time when we needed more JM-qualified peeps around. Of course, this only served to inflate my ego. Either way, my plan was to go to the promotion board, then to JM, and by the New Year of 2003, I would be Sergeant Kruger.

At the time, Army promotion boards had a format that hadn't changed in years. Four 1SGs and a sergeant major (SGM) asked numerous questions about the Army and military regulations to see how much knowledge you had and whether you were worthy of the next rank. They also asked about current affairs to see how aware you were of what was going on in the world and *blah, blah, blah*. We had to wear our dress uniform so they could see if we knew how to put that together correctly. The LRS 1SG aside, I probably had more schools, ribbons, awards, and pocket flair to line up on my uniform than most of the 1SGs in there. I had heard from some of the NCOs who had been in other units that when the infantry conducted a promotion board, and you had a Ranger tab, they asked you to say a stanza of the Ranger Creed, and told you, "Good job, you passed," maxed out the points you would receive, and sent you on your way.

LRS was under Military Intelligence (MI), and they all took turns asking me the standard three questions per three subjects per person. I passed. I would get promoted after attending the Primary Leadership Development Course (PLDC), which was affectionately named Pull-Dick from the acronym to describe what a waste of time it was. I would be allowed to attend after I completed the JM course.

JM was no joke. It was late September 2002, and three weeks of misery were upon me. My hatred for the other NCOs in my company was in full check; I used that as the fuel to drive me through. JM was playing the longest memory game I have ever played. You had to memorize the exact nomenclature of every part of a parachute, reserve parachute, rucksack, weapons case, and which side of the body they were on. You had to be able to spit out pre-jump verbatim without using any notes, and that thing was about a ten-minute spiel. You couldn't miss anything to pass.

While I was giving pre-jump to a Black Hat for a grade, I messed up a few sentences and would nervously chuckle at myself while correcting the misstep and go on. He tore me up over that!

"Being a JM is no laughing matter," he scolded me. "You better take it seriously, *Corporal!*"

I explained that I was laughing at myself for messing up in nervousness, but he didn't want to hear it. I added more hatred of him to my fire. I was relieved never to have to deal with him again. There were other rigging classes and tests to pass. Passing all of them with over 80 percent achievement would mean five chances at the Jump Master Personnel Inspection (JMPI) test.

The most challenging parts of JM were the JMPI test and Practical Work inside the Aircraft (PWAC). PWAC pertains to all the calls required to get jumpers up from their seats and out of the door safely and promptly. The hardest part of this test is getting the one-minute, thirty-second, and "Go" calls correctly. There were pictures of the locations to make these calls for each of the DZs on Bragg. The fail-safe was to get a correct one-minute call, then hum the *Jeopardy!* theme music in your head since that was exactly thirty seconds and then make the thirty-second call. "Go" was easy since you would be over the DZ then, and the green light would illuminate right next to your face.

JMPI, however, was a beast all its own. You had five minutes to conduct JMPI on three jumpers, two rigged for combat. We spent most of the three weeks learning JMPI in all dress conditions. I was glad it was not hot in October as we had to trade out wearing the gear for other students to practice. Half the class would wear the gear; the other half would conduct JMPI, and after an hour we would switch. Rucks were filled with lightweight cardboard padding, but it was still miserable to be out there with all this on while numerous students inspected the gear with you in it. JMPI required running your hands and fingers over every inch of webbing and straps to ensure nothing was misrouted, twisted, or improperly worn. Some people didn't have a gentle touch and would then be jerking you all over. I took note so I could return the favor.

I didn't pass the tests with high enough scores, so I only had three chances to pass JMPI: a pretest, the test itself, and a retest. I finished the pretest around 7:30 p.m. I thought I was moving out, too. How the hell would I take two and a half minutes off my time? When I took the second test, the Black Hat called "Time," as I was on the last jumper and failed. I had one more shot. It was not enough to inspect the jumpers. The Black Hats placed three deficiencies on the jumpers, and they must be caught *and* properly called out by nomenclature *and* where they were on the jumper.

It was October 28, 2002, retest day. I was eager to get this done, graduate, and move on. The next day, our company was going to Uwharrie National Forest, North Carolina, for three weeks of training, and I had to pack. *Focus.* I was relieved that the Black Hat administering my retest was one of the infantry guys in the pack of those non-grunt losers. My lineup was two combat-rigged jumpers, and a slick jumper. I knew I had to haul ass through the first two and could breeze by on the slick jumper.

"Begin!" the Black Hat yelled.

I moved out, started at the helmet, and checked it. *Missing a headband.* I called it out correctly, and since I knew I had the first deficiency and the only one for this jumper, I sped through him. *Done, jumper number two.* I kept my speed. *I'm on fire!* Helmet, harness, reserve, ruck—"Squat, hold!"—leg straps, turn, and on to the chute. I saw the deficiency—*an improperly assembled canopy release assembly, jumper's right side.* I don't remember what I said, but that wasn't it, and I snickered after I said it and moved on, hoping it wouldn't be a deal breaker. *Jumper two done, on to three.* I figured I had about two minutes as I was breezing through the first two, but I didn't slow down. Helmet, harness, on and on, parachute. I was almost done, no deficiency. *Fuck!* I missed it. *Oh well, I guess I'll get another shot on another day.* I reached back to smack the third jumper on the back of the leg to signal that I was finished and his JMPI was good to go. Right as I reached my hand back, I heard "Time!"

I was instantly deflated and numb, feeling that I failed and will have to return and go through the whole thing again. The Black Hat pulled jumper number two in front of me and said, "If you can correctly tell me what the deficiency is on this jumper, you will be cleared and graduate. I will point to it, give you ten seconds to think about it, and then give your answer. Ready?"

"Yes, Sergeant!" I replied.

He pointed to the deficiency I called out incorrectly, only gave me about two seconds to think about it, and said, "What is it?"

I smiled and called it correctly.

"Congratulations, Jumpmaster." He said, "Get out of here before I change my mind!"

I took my slip and ran out that door, the newest minted JM in the company. I still had to do PWAC since it had to be from an actual aircraft, and the day I was supposed to do it, we had inclement weather and were unable to finish it, but I knew the calls and all it would take was getting the timing correct, I wasn't worried. I'll spare the tension; I did pass PWAC later.

It was later that same day, on October 28, 2002, that I got a DUI—one of the most humbling experiences in my life. My anger and hatred for other NCOs turned to self-loathing and self-pity. I always held to the fact that the only thing I did wrong was get caught, as I knew plenty of other people who would drink and drive, too, though I knew it was something I shouldn't have been doing in the first place.

One of my fellow barracks dwellers had just turned twenty-one and wanted to see the glory of Fayetteville, North Carolina's nightlife. At least the bar nightlife that was around there. We had a half day, and I planned to drink with my roommate as I packed to prepare to go to Uwharrie National Forest for our training event. That's what I should have done. This newly drinking-aged man proceeded to beg me to take him out to the bars. I reluctantly agreed to take him if I could drive his Toyota 4-Runner, to which he agreed.

We headed out and went to the usual place: The Highlander. It was packed more than usual. We moved on to Huske Hardware, which was also slammed. After a few other stops, we ended up at BD's Mongolian BBQ, which usually had good music cranking in the bar area. The place was dead, and I was out of ideas for other bars to try. I still had to pack up for an early morning departure for training. He bought me a beer for my troubles, and we headed back to post.

Since 9/11, all military posts have gate guards. Air Force always had guards, but the Army started to enforce this after the fact. When I first got to Bragg, you could drive straight through without an ID; after 9/11, you had to have a military ID to get on post or get approved to be on post through a visitors' center, usually located at a main entrance gate. We were making our way on post, but there was a bunch of traffic, and they were checking everybody.

When we reached the checkpoint, the military police (MP) said, "Do you know your registration has expired?"

"It's his car," I pointed to my buddy, who, by the way, was trashed.

The MP walked to the other side of the ride and asked him for his license, registration, and whatever else. As he was rifling sloppily through the glovebox, the MP asked, "Have you guys been drinking?"

Now, at this point, some quick-thinking individual may say something like, "I have, and he came to pick me up," or some trouble-reducing statement like that, but no, not this guy.

He looked at the MP and said, "Yeah."

It was like the snap of a finger, and the MP was back at my window with a breathalyzer saying, "Blow into this."

I blew a .14. I was such an alcoholic back then that I didn't even really feel fazed by a .14. As far as I was concerned, I felt the same as if I'd had a couple of beers. They pulled me out and, right in front of all that returning Friday night traffic, gave me a field sobriety test. I thought I did well, but when I read the report later, I saw I hadn't been the twinkle toes I thought I'd been. Don't get me wrong, I didn't fall or say some dumb shit like people do, I just didn't ace the test. The MP read me my rights as he cuffed me and put me in the Ford Explorer. Yes, I thought about running since I had access to the front seat, but they still had my ID card, so they could have found me too quickly. I sat back and was ready to endure the consequences of my decisions.

We arrived at the MP station. They took the cuffs off and placed me in the holding cell, which was a small room with a double-sided window. There was a guy in there who was way drunker than I was, and a mouse on the floor that was scurrying around what looked like a small puddle of vomit. I hadn't been combative, as most drunks can be when they are getting arrested, so they had mercy on me and moved me to a chair next to some of the other office cubicles. This time, they took another breathalyzer with a full-size machine, and it was .16. That would be the final number used in the report. Battalion Command came and picked me up and escorted me back to the barracks. Within minutes, my phone rang.

"Kruger! DID YOU GET A DUI?" my PSG blasted through the phone.

"Yes, Sergeant," I said quietly, sheepishly.

"I bet you feel pretty sober right now, don't you." His tone was a little less harsh but still excited.

"Roger, Sergeant." The thing you say in the Army when you don't know what else to say.

"We'll talk more about it in the morning. Make sure you're packed and ready to go on time," he finished.

"Roger, Sergeant."

Damn, hypocrite! He was probably out doing the same thing.

I called my parents, who were three hours behind, and told them what had happened.

"Hey, Chris! Everything OK? You don't normally call this late on a Friday night." My dad said when he answered the phone.

"I just wanted to call and let you guys know that I just got a DUI," I informed them through a lump in my throat and holding back tears—the emotions of what had transpired finally catching up to me.

"Oh . . ."—there was a long pause—"Is there anything we can do?" My mom was now on the phone with him.

"No. I don't know what will come of this or if it will end my career, but I wanted you guys to know what happened."

"Thank you for being honest. Let us know if we can do anything." An offering that was more of a consolation than anything since I knew there was nothing they *could* do for me.

"Will do. Love you guys. Talk soon after the training is over." I ended the call.

I packed up and got ready for the events of the next three weeks, but I didn't get any sleep pondering what my future would be.

CHAPTER FOUR

Forty-five days of restriction to post; forty-five days of extra duty; the loss of half a month's pay for two months; the reduction in rank from corporal to specialist; being flagged from promotion for six months; and the one that hurt most of all, no favorable action. This was the Article 15, a non-judicial form of punishment for my DUI. I wouldn't be court martialed or kicked out of the Army for it.

No favorable action meant I was not able to take Christmas leave, which pissed me off immensely. I found out while I was on extra duty that we were deploying to Iraq in the first week of February. That information came along during Christmas leave, and the command team deliberated on whether to initiate a recall to bring back all the soldiers from their leave early. They opted not to since there would be enough time to accomplish all of the pre-deployment preparations when everyone returned from leave. I inquired whether I could take some leave after my forty-five days of restriction and extra duty, but the answers I got were akin to Mom and Dad saying, "We'll see about it." I'll save the suspense; I didn't. I never got a real good reason why, either. Oh well, mark that as the longest I've been away from home without a visit. It would be nearly another year until I could see family again. I was somewhat devastated and somewhat bitter, but mostly mad at myself and full of self-loathing because I always owned it as my fault. Hatred and drunkenness became close companions over the next months.

Our deployment date got pushed past the Super Bowl so we could stay and enjoy one more before we deployed. I would have probably been alone in self-pity if not for faking being cool about things in front of the others.

We arrived in Kuwait on February 10, 2003. I saw the flat, desolate, packed sand go on for miles and wondered how there was still so much here: roads, cars, and all this in a bland, dull area. Here we were on a bus, packed nut to butt and having to hold a piss for an hour and a half straight. It was early morning when we arrived at Camp Virginia, and it was already getting warm. I couldn't believe how hot it was in February, which made me a bit concerned about how stifling hot the summer would be.

As a LRS unit, we had to validate that we were, in fact, able to accomplish our mission with full load to the standard we laid out in our standing operating procedure. We had to infill with a full combat load of thirty-five or forty pounds, ruck with 120 pounds of gear, move twelve miles in one night, set up for three-day surveillance, and extract another five miles or so on the third day. The only saving grace was that some weight was water and food, so the load felt slightly lighter on the way out. We set out at 10 p.m. when it was finally dark, and after a quick stop to check out some lights we saw, we made it to the team HS twelve miles later, about thirty minutes before sunup. Rockin' it! Those lights, you may ask? Marine force recon was in the area conducting long-range communications training. Two Marines were out there, and both focused on the radio issue. We snuck up on them and spooked them a little. They probably put a little more stock into security after that.

We only pulled one day of surveillance, which was good because all that was there to watch was flat open desert and a couple of old tank ditches from the Desert Storm era, or so we were told. We were to be exfiled by a Black Hawk helicopter, but incoming messages showed we were to move another five kilometers to ground extraction.

"That's bullshit!" the TL blasted as he typed a message back to the CP. We all agreed.

"We're not walking anywhere. Send the Hawk." He typed into the laptop and hit send.

"Bad weather, nothing flying. Move to the new extraction grid." The reply.

We pitched a massive fit over that as we didn't want to move, let alone ride the whole way back in a 5-ton, which is what we did. The drive felt like it would never end. We found out later that a Black Hawk crashed while conducting training, so they grounded all birds in the area of operations (AO) due to the dust storms. The dust storms made the ride in the back of the 5-ton worse. If you close the tarp in the back, it's hard to breathe. Leave it open and be covered in dust, which also makes it hard to breathe. We closed it up and dealt with it that way.

We conducted other training during our five weeks at Camp Virginia: hide site construction in a desert environment, long-range comms, and physical exercise to help acclimate to the environment. These preparations became common to all the deployments. We would be in Kuwait for a time, train up on whatever was necessary, move into Iraq and operate for the deployment's length, return to Kuwait for a few weeks, and go home. It became as regular as breathing, but not quite yet. It was new to us; we were at war on terror, and my journey into this realm was beginning. I knew the obstacles ahead, which were simultaneously thrilling and terrifying. Were we going to get into firefights? I didn't know, but I hoped so. We had just watched the Super Bowl, and now we were headed into a Super Bowl of our own. The Big Game. *It's time to shine, put up, or shut up.* With my anger sorely focused on redemption from my slip-up, I was ready to prove I was still trustworthy and had what it took to be an outstanding NCO. Fortunately, my chance was on the horizon.

Two-and-a-half years of training, and this was it, war. *It's on!* I think some of us felt that as soon as we breached that berm it would be bullets, rockets, and any other type of incoming fire, like the random shooting in the sky I remembered from the Desert Storm footage we saw on the news only twelve years prior, give or take. As convoy after convoy slowly rolled through from Kuwait into Iraq, it was more like a long, slow road trip through a hard-packed desert. There were roads here and there, but those were primarily avoided to create our way through the dry, desolate wasteland. We would see many blown-up vehicles along the way and identify them from our "Vehicle Recee," which was part of our daily, non-physical training. A lot of old tanks from the '50s or '60s, personnel carriers, and some fighter jets here and there that were stripped for parts. *Who could win a war with this dated crap? Let alone stand a chance to fight without air superiority?* We would learn that where there's a will, there's a way.

Then it happened. The light moving tactical vehicle (LMTV) that I was driving died. We had mechanics with us, so we halted somewhere in the middle of nowhere. Whatever was wrong with it couldn't be fixed with what we had on hand, so we had to be towed. It wasn't long before the other two riding with me and I were asleep while being towed. I slept leaning over the steering wheel, mostly keeping the LMTV on track until the tow vehicle took one turn too sharp. As they were straightening out, the wheels of my ride stayed turned and dragged them to a stop. They thought their LMTV was breaking down too. I told them I couldn't see through the dust and turned the wrong way. I apologized for the error and laughed, but I was laughing because we were all asleep. As soon as we took off again, we were all back out, and this time, the vehicle stayed the course with me racked out on the steering wheel, with no helmet and no care in the world. War, wow.

We met up with and passed convoy after convoy. We would get fuel when we met with the trucks that had it. Fortunately, it was March and not very hot yet as "they" wanted us to wear J-LIST while we traveled. I forget what that acronym stands for, but it was the chemical suit we wore. The tops and bottoms of the suit would protect us in the event of a biological or chemical attack. Nuclear, though? Forget about it. This thin suit lined with charcoal wasn't saving us from that. I didn't mind wearing it since it was warm enough to wear and sleep overnight without overheating or getting cold. Plus, we could pop up at a moment's notice and be ready to go. Most of us would find some loose sand, lay down, nuzzle our bodies into it, and sleep. It was like making sand angels, but with no wings.

On one of the days that we were rolling up to Objective (OBJ) Rams, our primary destination from day one, we rolled past a firefight. We were still being towed at this point and were looking at the city off in the distance. We watched as some A-10 Warthogs were maneuvering to avoid ground fire. Then I saw a bright object headed toward one as the A-10 flew nearly vertically. I don't recall if the A-10 launched flares, but the object following it missed; the A-10 rolled around, was headed nearly straight toward the ground, and then swooped back up to level flight about 500 to 1,000 feet off the ground. As I was about to look away to see if anything was going on anywhere else, there was a massive explosion. I assumed that's what a 1,000-pound bomb looked like going off, and the A-10 received no more ground fire from that location. *Got 'em!*

Almost simultaneously, a bunch of soldiers lined a berm we were slowly driving by. They were shooting off in the same direction as the city, but there had to be at least a mile of desert between us and the city. I didn't know what they were shooting at since we saw nothing worth shooting at in the flat, beige landscape. That didn't stop one of them from yelling at us.

"Put your gear on. This ain't no weekend drill!"

We laughed at the notion that he thought we were National Guardsmen at war, but still, we couldn't see any incoming fire. The thing about helmets was that LRS is long-range surveillance, so helmets were unnecessary weight to carry, and for us, airborne-only items. Little did I know we would shortly become much better acquainted with wearing them. Goodbye hair, not that it mattered. I was balding anyway, and I knew it. I embraced it when I was young, so I didn't care.

Fake firefights aside, we made it to OBJ Rams. We set up some of our tents and waited for the missions to roll in, and they did. Airfield security for the drones launching at a tiny airstrip within the objective area. *What. The. Fuck!* The premiere LRS unit in the Army, supposed to infiltrate ahead of the forward line of troops (FLOT), fearless in the presence of a surrounding enemy force, relentless in getting to the OBJ and reporting timely and accurately so that the next friendly forces engagement went off without a hitch. We were degraded to guarding some small strip of concrete so that unmanned aerial systems (UAS) could launch and gather the intel that our "eyes on" were supposed to get. We were deflated, big time.

MREs were the meals of the day. I got so sick of them after a couple of weeks that I would eat one a day—Ranger

School flashbacks. I wasn't hungry; I wouldn't say I liked eating them non-stop. After arriving at OBJ Rams, we routinely received T-Rations, or T-Rats as they were called, a welcomed change from MREs. We would boil water in a large trash can and drop the large packages they came in into the water to heat them. They weren't much better than MREs, but they did have breakfast varieties.

During that time, we had some usual infantry shenanigans. We started to use the water buffalo, a large cylinder to hold water attached to a trailer, and shallow bins meant for washing our clothes, as our bathing system. We would fill the shallow bins with water, get naked, and bathe in them. Seeing a few guys doing that at a time wasn't unusual. Of course, we had our security set in place too; we weren't complete slugs. The NBC (nuclear, biological, chemical) alarm, three even horn blasts, would go off regularly. We would have to put on our J-LIST and mask and wait for the all-clear signal.

One guy would not get up for the alarm or wear any gear. We would watch him for five minutes, and if he wasn't twitching, we would all take ours off and hang out in the tent until the All-Clear was given. We called it the "All-Clear Tent." Bored to tears, one day, someone had a ball, and we took a cot pole and played stickball until our CO yelled at us for having fun in wartime. We had makeshift gym equipment with sandbags, plastic chairs with holes cut in them so we could sit and shit in comfort, and sometimes a few of us would sleep in the LMTVs on top of everyone's duffle bags since it was just a bit more comfortable than a cot. Lastly, we all tried to be gourmet chefs with MREs and make something that tasted great.

One night, we were adding some of the cleaners to the water buffalo, and while standing on the wheel ledge, a massive explosion went off, followed by the incoming alarm.

We got used to donning our equipment quickly, so we ran for our gear. The blast was so big and bright that I could see my buddy's face across the water buffalo plain as day for about two seconds. My mind raced and I thought, *Nuclear?* Then the glow disappeared, and we saw where the blast had come from. One of the Apaches had launched a dud Hellfire missile, and Explosive Ordinance Disposal (EOD) went to finish it off. No one announced it over the radio, so it caused panic for a couple of minutes.

Then it happened: the biggest sandstorm any of us had ever seen. It lasted for three days—for three days we could barely see fifteen feet in front of us. At night we could see five feet ahead at most. It was red earth, looking literally like movies about Mars. It was so bad it halted the war for everyone involved. I'm still digging that sand out of my orifices over twenty years later. In 2016, while I was at the Advanced Warrant Officers Aviation Course (AWOAC), I heard an attack company CO's account of the ordeal. The storm had come right after a mission during which one Apache was shot down, with the crew surviving and being captured, and a bunch of other Apaches taking fire and needing repair. For them, the sandstorm was a godsend to a relatively unsuccessful mission; to us, it was a nuisance, and we couldn't wait for it to end. Oh, they recovered the captured crew a couple of weeks later, and I also got to hear their story in SERE (Survive, Escape Resist, Evade) School.

Dust storms finished, and after a couple more weeks of babysitting the drone folks, we were called to join the battle; we were going to Baghdad, Iraq. As we entered Baghdad, we drove under a monument of hands holding crossed swords. It looked cool but left me wondering what it signified. I quickly dismissed the thought while driving and focused on road threats. I would come across numerous road decorations such as this one—elaborately tiled road coverings that looked out of place, connecting to nothing and in odd locations. Their placement made it seem as if they'd been built under the thought, *You know what would look great right here? Some decoration for no reason in an obscure location that no one will appreciate.* The ornateness of some of these decorations strangely juxtaposed their bleak surroundings.

When we arrived in Baghdad, it was honestly pretty trashy. Sure, it's a big city rich with history, whether good or bad, dating back to its empires of ancient history, but I had figured it would be somewhat nicer. In reality it looked like everything had been built on cinder blocks and concrete in a slightly unstable fashion. We set up shop for the night in what we called Sad Disney. It was a large open area surrounded by an eight-foot-tall cinder block wall with run-down amusement rides and pictures of Mickey Mouse and some other Disney characters poorly painted on the walls. The rides were the type seen in the kiddy section of a theme park. They all looked as though they'd been broken, rusted, and inoperable for a long time; who knows? Even the trees looked sad as they swayed gently in the hot breeze, covered in years of dust. Their greenness had lost its luster having constantly been layered in tan dust. Surrounding our makeshift compound were taller buildings that from above made Sad Disney look more like a shooting gallery from above than a secure area. *Come shoot at the American soldiers who have no body armor or up-armored vehicles to retreat to. Throw a grenade over the wall and hope you splatter a couple more.* We stayed there for the night, but the battle was moving so quickly that we moved on to Tikrit come morning.

On the way to Tikrit, I decided I wanted to be a pilot. We had OH-58 Kiowa Warrior helicopters flying convoy security and recon ahead of the route as we drove. One main highway spanned the length of northern and southern Iraq. Any guesses as to the name of the highway? *Shut up, you prior service dude in the back; let the POGs guess.* That's right, Highway 1. *So original,* Saddam (insert eye-roll emoji). As I was driving, which I did for nearly the entire deployment, I would watch them fly along at thirty to fifty feet above the ground, swoop left or right to check out a building within small arms range of the convoy, buzz by the buildings, and move on to another. They would stay Nap of the Earth (NOE), within 25 to 100 feet off the ground the whole time. On occasion, when they would see something of interest, maybe a potential target, they'd climb at an angle with the nose of the helicopter pointed skyward, conduct what looked like a U-turn, and dive toward whatever they were looking at to see it straight on. When it looked like they were about to crash, they would swoop back up, or level out and continue. It looked like such a blast. I knew I wanted to fly one. They also flew doors off, which seemed to be a more comfortable way to travel than in an LMTV with doors and no AC.

We arrived at a secured airfield in Tikrit, where things seemed relatively peaceful. I thought we'd missed the war since we continued to arrive at secure locations with no resistance along the way. Or that it was all a training joke. Someone tranq'd us on a plane and we slept, arrived in a desert with a bunch of actors who were good at being Middle Eastern, and let the training commence. We were in Tikrit for a day or two, and then we headed to Mosul. Surely, there would be some work for us in Mosul. Sometime later, I learned Mosul was the location of Nineveh in the Bible. When the great fish spat out Jonah, he went to Nineveh to warn them, "Forty days, and you're done." That was modern-day Mosul. Well, the northeastern side of what is now Mosul anyway. You can Google it.

It had to be the beginning of May 2003, and we had finally arrived at the battlefront, or so I thought. We had a hardstand building inside a secure location with some amenities around.

There were MKTs (military kitchen trailers), so MREs were no longer the sole meal option. Wooden-built porta-johns were a sight, and we set up stalls to hang solar showers. We received care packages, a nice change from relying on everything we carried. We could go beyond our cinder block prison and engage the locals for food and comfort items. Some guys got sick from eating the local cuisine, so in true infantry fashion, we made fun of them for having weak stomachs, or weak genes, as we would continue trying some of the local foods. After a couple of months of MREs, we took the chance of getting sick to eat something different than the usual crap. Chicken Tikka was pretty darn good, and when they cut that meat fresh off the rotating stick, it was phenomenal. Chai was a bit different than in the United States as well. In Iraq, Chai is made in a big pot with loose-leaf tea. Then, it is served in a shot glass with so much sugar a diabetic would cringe. We joked with a few of the locals that we were adults and wanted it in an adult glass. They thought it was weird but would oblige us, though it never caught on as a local trend.

The Tigris was a little disappointing. I figured it would be this gigantic river gleaming beautifully and shimmering with the sun shining off its glorious waters. Maybe it was broader in other areas, but it was dingy and had a green color that neither shimmered nor shined. It looked as if the tree I described from Baghdad had been liquified, sad and infected from years of neglect and abuse. It was still rather awe-inspiring to be driving on a bridge over waters of what seemed to be folklore. Majestic? Not really. Something noteworthy, though, as most Americans only know it by name.

Finally, missions were rolling in. Mosul had yet to be a hotbed for combat like Fallujah and other areas. At this point we were so wrapped up in our expedition that I didn't know about Fallujah yet or the aviation ordeal I described earlier. I figured everyone was having an easy go, so I thought the threat level was nonexistent and everyone was doing great. I still had no idea that people had been captured there as prisoners of war (POWs), that some had been killed in action (KIA), or that the Marines had fought a massive battle there. I knew of Fallujah only because I'd seen it on the map. I wasn't ignorant; it just wasn't our fight. We were about to discover our battle, but I'll save some suspense; it was a battle against boredom, mostly.

The company's mission set picked up right around the same time we had heard that President Bush would call an end to major armed conflict. Rumor had it that our time in Iraq would be cut short. False motivation that would start to tear at people's emotions. In the meantime, our team received our first mission. *About damn time!* We were to set up a SS on the edge of Mosul, surveil Highway 1, and report anything suspicious traveling into Mosul. There were already reports of insurgent fighters entering Iraq to help the Hussein regime stay in power or take a shot at some Americans to deter our way of life from entering their country or region of the world. Three-day mission, report suspicious activity, no problem.

We would of course infil at night by Humvee, about ten clicks from where we lived. The terrain was lightly rolling hills, and illum was next to zero until the sliver of a moon came up around 10:30 pm (No, I will no longer use military time; I'm retired and am glad to leave that crap in the past). We loaded up with extra ammo and every explosive we could: thirteen 30-round magazines apiece, grenades, AT-4 rocket launchers, and Claymore mines. Our team would be at one site for extra security—six dudes versus the world. We walked up one hill and down the next. The NVGs worked great in the limited light, and we were as sure-footed as goats on the gentle but rocky terrain. Arriving at a saddle in between two hills, we took a rest.

As I pulled rear security, I saw a slight gleam of light behind us at the top of the hill in the direction we had just come from. *Someone is following us,* I thought to myself. I kept an eye on it but had yet to alert the team. The light was getting a bit bigger but very slowly. It got larger, and I thought it was a slow-moving vehicle. I alerted the ATL as it continued to get bigger.

"Hey man, I think I see something."

"What is it?"

Finally, the light got big enough, and I realized the moon was finally cresting the horizon. "Never mind. I was watching this light, but it's just the moon."

"Dumbass," he said, as we both started to chuckle.

We arrived at our SS, and there was a flat area with tons of large rocks around, near the gently sloping top of a hill, so we set up in the midst of them. We placed the Claymore mines in the most likely avenues of approach along with the AT-4s and a couple of the grenades. If anyone wanted a fight, they would get one for sure. We set up comms so we were ready to report any devious activity in real-time. It was an hour or two before sunrise, so we pulled security with a two-man watch so the others could sleep. After sunrise, we watched the road and saw nothing particularly alarming. It appeared as regular morning rush hour traffic, nothing to write home about.

Around 9 a.m., we received a message asking why we weren't reporting anything, and we informed "higher" that there was nothing that fit the description of what we were to report. They insisted that we report the activity we saw, so we sent traffic reports non-stop for the next hour or so. We were their morning news traffic copter's eye in the sky, but we sent them every suspicious vehicle instead of giving an update on their commute. A Bongo truck with a bunch of propane tanks on it? Sent it up. POS car that looked rigged for a bomb? Sent it. Dude walking to work? Ding, sent! "Higher" finally came back and said something to the effect of, "Point taken, resume surveillance and pass only suspicious vehicles and activity." Idiots.

Other mission sets were rolling down, so we only stayed there for thirty-six hours. They cared so little about the mission that they drove straight to us and picked us up midmorning. Tactical. Returning to the compound, we learned that one of the other teams had been compromised and engaged in a brief firefight as the enemy fled. Another team had passed up some actionable intel, and an Apache destroyed a building. That's what we wanted: to be in the game for real. To be the cause of some destruction or action that was noteworthy. Nobody was looking for a Silver Star, but everybody wanted to be the hero of the day, the reason real shit happened or got done. Like the superhero in a movie who saved the day with one final victorious action. This was where it would happen. Don't get me wrong, I don't think most people set out to make something like this happen deliberately. Responding to a situation courageously leads to the outcome most people seek. If I had to guess why someone was dissatisfied with their service, it's because it never reached the greatness it could have unveiled. A park ranger saves a lost family, a mall cop foils a robbery, the mailman saves an accident victim, and a dull job unveils greatness. Our chances for glory would significantly increase as we began patrolling Mosul regularly, and we relocated our company to a small facility in the heart of Mosul.

There was a three-story, cinder block, and concrete building where we set up shop. Our team had a room that was a good size for the six of us. We each had room for a cot and personal stuff to the left and right, with a little area left for the coffee urn we brought for the whole company to help themselves. I would intentionally make the coffee too strong, which we all liked, so those with weak genes wouldn't come by because the strength upset their delicate tummies.

At last, we had real bathrooms. We acquired some hoses and made a couple of outdoor shower stalls. The water was cold, but as it was now June, the temperatures were already hitting 100 degrees, if not over. No one cared about the water temperature, even if we showered at night. We were on our own and had our areas and spaces; we set up building security and conducted patrols every other day by vehicle. War life was pretty good.

Care packages and letters came in regularly. Among numerous letters and care packages I received, my Great Aunt Audrey wrote me almost more than I ever expected. She was in her early sixties and hilarious as she would write things like, "If you don't like the goodies, then just give them to someone else. They'll think you're being nice and will never know you didn't like them!" One time, she sent a slingshot and marble-sized jawbreakers and wrote, "If you hit kids with these, at least you're giving them a treat too!" She would finish every letter with "I love you! It's meee, Great Aunt Audrey!" Which would always make me smile. Yes, there were three Es every time.

When the war kicked off, technology wasn't what it is today. It was at least a two-week wait if you didn't have something you didn't bring or wanted to get while deployed. Amazon was getting off the ground, so friends and family sent items through care packages. The internet was not as high speed as it is today in the States and was even slower in the Middle East. There was no FaceTime or video calling. We only made phone calls when we had time to get to a phone, and the phone trailers of later deployments and the use of minute-loaded phone cards weren't much of a thing yet. We had to wait turns to use a computer to send an email, and if you wanted to wait to use a free phone, the wait was usually long, and the allotted time was five or ten minutes. It was a soul-crushing disappointment if you called someone and they did not answer. The camaraderie was high, though, as we did everything together: watching movies, listening to music, playing board or video games, eating a meal, everything. It broke down barriers, and you got to know someone inside and out. We shared everything, good times and bad times.

On June 4, 2003, I experienced one of those bad times. I had a step-grandpa on my mom's side of the family who passed away when I was eight years old, but I knew him as my real grandpa since he was the one married to my grandma and the one who was always there. There was another grandpa we called Grandpa Joe. Later in life, I found out he was my real (biological) grandpa. He traveled with his wife in a fifth wheel and would come to visit for a week or two every couple of years, so I had no relationship with him but no animosity either.

"Kruger, come see me when you get refit from the patrol," my PSG said. We had just returned from patrolling the northwestern side of Mosul.

"Will do, Sergeant," I replied slowly, perplexed. I had no idea what to expect.

After dusting off my M4 and resetting my gear for the next mission, I went to the PSG. "What's going on, Sergeant?" I asked quizzically.

"There's no easy way to lay it out, so I'll just say it straight. We just received a Red Cross message, and your grandfather has passed away."

After hearing that information, I think most people would be numb, lightheaded, or a bit limp legged. I assumed he was close to his grandparents, and this news would have been more shocking to him than it was to me.

"Oh, OK," I said. I was processing the information, but I was indifferent to it because there was nothing I could do with it.

"Were you close with your grandfather? Was this expected? You don't seem very surprised."

Sparing him the story mentioned above, I simply said, "I knew he wasn't in great health and had been battling cancer, but no, we weren't that close."

"OK. We'll see what we can do about sending you home for the funeral, but with everything going the way it is, I wouldn't count on it. Let me know if you need to call your family or talk to someone about any of it, and we can make it happen." A consolation prize.

"Roger, Sergeant." The thing to say . . .

They told me that I would not be able to go home for the funeral, you know, because of war and stuff. Death in the family? War first, even though the president had announced the end to major combat in Iraq.

So, back to the mission. Patrols were somewhat fun at first. Rolling around the west side of Mosul, we quickly learned the area, and it was good to be doing something other than just waiting. Looking back, we all should have been more cautious as we were riding in a Humvee with no armor and a sheet of plywood across the roof where the driver and passenger were so that our SAW gunner had a stable platform. If that doesn't scream, "Take a shot at me," I don't know what does. An area in the center of our patrol sector was a massive garbage heap that took up a square city block. That smell is ingrained in my nostrils and memory forever. They say smells can take you back to places. When I smell something akin to that combination of trash, B.O., shit, and stale air, I'm immediately taken back to being a rolling target in Iraq, not in a "dive for the bushes" type of way but in a disgusted with the smell type. Seeing women wearing black burkas in the middle of a hot day was always weird. I understand its cultural significance, but man, have some sympathy. Plus, we never knew if it was a man wearing a suicide vest (S-vest) or not. I took a personal posture of hoping for the best and expecting the worst.

The patrols continued for a month, and then we got an interesting mission. They wanted ten guys to fly up to the mountains in northeastern Iraq, near the Iranian border, and be the first to meet with the Peshmerga soldiers who lived and operated there. We learned that they were some badass Kurdish fighters who had fought off Hussein's regime several times and had unofficially declared northern Iraq as their land. They would refer to the land as Kurdistan, though it was never officially recognized. They added me to one of the other teams to go and be the first to shake hands with these fine fighters and be part of the beginnings of forming or rebuilding a shattered alliance.

I never read the history books on this but was informed that during Desert Storm, the United States allied with the Kurds to conduct a pincer move on Hussein, but after four days of that battle, we pulled out, leaving them high and dry, so our relationship with them had been in tatters since. *That sounded about right.*

The Black Hawk we flew in on landed on a peak with a small outcropping, and there were three Toyota Hilux trucks (think, Toyota Tacoma but diesel) to pick us all up. We drove down the beautiful hillside to their camp, which wasn't much more than a handful of buildings, some cinder block, but most were hard-packed mud and straw brick. Seeing such beauty in a place that seemed only to have so much desolation as we flew up to these mountains was amazing. The greenery was more vital and less dust-infested than it was in the rest of the country. The mountain air was still dry like the desert area we had just left, but it had a coolness that had departed the lower areas months ago. It was a sight to see; it was beautiful, serene, and quiet.

The first order of business was lunch. We watched as some soldiers took a goat behind the building we were in, which I thought nothing of, and then, about an hour later, brought rice, kabobs, and some veggies around. Curious, I asked what we were eating, but not because I'm picky. They snickered while our interpreter said, "Didn't you see the goat?" I smiled back, raised the kabob in a toasting manner, and dug in. Fresh killed and cooked goat kabob was fantastic! I'm not sure what they put in the rice, but that was good stuff, too. They probably used the goat leftovers to boil it in so it would soak the flavor—ignorant bliss in delicious food. I also liked the salad, which was nothing more than shredded cucumbers and tomatoes with oil, salt, and vinegar.

After lunch, the adults went to talk. I was mainly part of the security, so I never really got the details of the conversations or what came of it. I enjoyed seeing a new area that felt like a completely new country. They drove us down to an Iranian checkpoint and showed us the tanks just beyond on the Iran side. We had a couple of fly-bys with fighter jets, and a couple of Apaches that were in the area with us flew a little too close to the border, and the Iranian tanks moved positions and began to orient their guns in our direction. There's nothing like almost inadvertently causing an international incident with what was supposed to be a peaceful meet-and-greet.

Interestingly, these fierce fighters appeared to be regular dudes like the rest of us. I had this image built up that they would look like super soldiers. Jacked dudes out here running up the mountainside. Nothing of the kind. They looked like regular guys, but I did not doubt they would run circles around us in those mountains if push came to shove.

With the meet and greet complete, we headed back to Mosul. We made routine trips to Dohuk, a Kurdish-controlled area in the north just beyond the first mountain ridges of Iraq. I tried to get on these trips since the mountain areas were always slightly cooler and easier on the eyes. The Kurds maintained their land better than most of the southern regions as they seemed to better keep up their infrastructure. Again, it was more handshakes and conversations I wasn't privy to, but the food was a welcome change from our usual.

A large group of us were in Dohuk on July 22 when Uday and Qusay Hussein were killed. Everywhere, the locals were celebrating. I learned from one of our interpreters that the Husseins were responsible for some chemical bombings in the Kurdish areas in the late '80s, so there was apparent hatred from then, amongst other reasons. Learning of Hussein's sons' deaths caused them to celebrate. Drinking alcohol is forbidden in Islamic culture, but the Kurds didn't adhere to this as they had liquor stores in their regions. As we were walking down one of the streets that day, a drunk Kurd was shouting praises, "America, OK! Uday, Qusay, no way!" That may have been the extent of his English. As he was saying this repeatedly, he grabbed my shoulders, kissed me on the cheek with an "OK!" and went on his way. We all laughed, I shook my head, and we went about our business.

Around this time, we experienced our first combat losses that were *not* combat-related. Our HQ was in Tikrit, and they had a change of command (CoC) that required maximum participation. Our company was to send twenty personnel by vehicle from Mosul to Tikrit to be present at this CoC. Two Humvees and an LMTV hit the road for the two-to-three-hour drive with more than enough people for security and presence at this phenomenal waste of time and resources. En route, the convoy ran into bad weather, and the LMTV driver was relatively inexperienced. They didn't see the cratered-out portion of the road before them until it was too late. The driver swerved to avoid the enormous pit in the road, but there was no way to prevent the ensuing rollover at the speed he was going.

There were some Air Force loading pallets in the back on the floor that crushed two of the passengers; they died instantly. The other guy was catapulted face-first into one of the overhead tarp crossbeam supports, busting out his teeth only held in by his braces. He would show the same braces when asked, "Why did you join the Army?" One of our guys was a prior Ranger Battalion-dude-turned-civilian-EMT who came back in because he missed the military life. Camaraderie goes a long way, especially in the infantry realm. He was able to help keep that injured soldier alive until medevac arrived. It took the medevac Black Hawk over an hour to reach them as they had to wait for good enough weather conditions to launch. In the end, he passed away, too, after getting into higher care. It took too long to get him there, and the lung injury he sustained in the rollover was the cause of his demise.

"I can't believe it, man. Gone. Three good dudes. Just gone. And for what? A fucking change of command!" I was sitting on the roof of our building, trying to process with Nate what just happened. Even though he was a commo guy, we made quick friends as he was from Walla Walla, Washington, and we had a geographical connection.

"I know, man . . ." and then silence, his reply.

"Why the hell did they need more guys there anyway? They have the rest of the battalion down there in Tikrit. They needed twenty more for the dog and pony show?" I blasted rhetorically.

"This is some bullshit!" Nate sneered.

"I thought we were at war! I couldn't go home for a funeral but putting twenty guys at risk for a fucking change of command is OK?" This is where my disdain for wastes of time like this has grown. I understand having max pax (pax being short for personnel) at ceremonies and events in the Army is necessary to continue in traditions and ceremonies. However, having twenty more dudes at a CoC in a time of combat that contributed to the loss of life and manpower was negligence at best. Do you think I'm being unfair in that assessment? Maybe. Who's to say those guys wouldn't be alive today doing great things in the world or with their families? Dead, due to a fucking CoC!

Anyway, we had one other short-lived opportunity for a cool mission. We were to be on twenty-four-hour alert to launch and interdict suspected terrorist training camps in the desert areas in western Iraq. Our team and one other team relocated to Q-West (I have no idea what the Q stood for or why it had that name), where most of the aviation assets were located. We trained with some CH-47 Chinook crews and had to be ready to launch immediately. We mainly worked on our comms equipment and practiced expediting the loading and unloading of a Nissan Patrol from a Chinook. Backing a manual transmission vehicle into a Chinook was challenging as it has only inches on each side until you're causing some damage. Still, I managed to drive that thing up the ramp with minimal issues and no damage to the vehicle or the airframe.

We wanted to land with the team loaded up in the vehicle, drop the ramp, and roll off within thirty seconds to one minute of landing. While the lead pilot stated that would be awesome, he also added, "We're not the 160th, so we'll have to slow it down and let the crew chief direct us out." In the end, it would have only taken two minutes to get the vehicle out, load the team, and go, but that's still an eternity to leave a signature with hurricane-force winds flapping on the ground. The mission came to nothing but a fun couple of weeks of training, a few helicopter rides, and training with Kiowas on aerial calls for fire. Then, we were back to the actual work.

Around this time in the summer, my flag was lifted, and I attended a relatively easy promotion board. The 1SGs and command sergeant major (CSM) were all impressed with my Enlisted Record Brief (ERB) and all that I had accomplished in three years, so they only asked me a few questions like, "What did you think of JM School at Bragg?" and "At what point did you want to quit Ranger School because everybody does at some point?" Perhaps the lamest of all was, "What does LRS stand for anyway?" Compared to my first promotion board, I was in there for four minutes at most. They gave me maximum points, said, "Congrats, Sergeant," and sent me on my way. My DUI didn't set me back at all. I was promoted along with my peers, with whom I arrived at Fox Company.

We were thrown a tasty bone in late summer 2003. Special Forces (SF) didn't want a particular mission and mentioned that the LRS guys may be interested. That's what we heard about it, anyway. Whether it was true or not was irrelevant. We were to go to Zakho in the north-western corner of Iraq near the Turkey border and begin training the Iraqi Border Patrol (IBP). Zakho had a Kurdish Officer Training Academy, which we would occupy, and it was our job to train them to keep their northern border secure. Also, we would have funding to set up whatever we needed to facilitate this mission. When we got to the training center they used, we were ecstatic to find that there was already a twelve-foot wall surrounding the whole compound, an actual gym (though not a great gym), hardstand buildings that had usable indoor plumbing, and most importantly, a pool. It was a nice pool with three levels of diving boards. The temperatures were easier to deal with since it was in cooler mountainous terrain, and the views were spectacular as far as the eye could see. Green was actually green, not dirt-covered green. Zakho had another thing that got us in some hot water: beer stores.

Our first order of business was the usual "foxhole improvement." We ordered and received air conditioning for the sleeping and operations areas so quickly my head spun. We had beds again, actual beds! I had a twin-size bed and wall locker all to myself. It was a bit like basic training, still in a large bay, but it was nice not to be on a cot or a wooden bunk bed with a four-inch foam mat. We had a shooting range behind the compound that was nothing more than six-foot-tall berms and wooden target stands, all facing in a direction without a house beyond the berm. We got all the trainees new gear, but they were primarily interested in the boots. All their equipment was worn out or broken, and they were just as happy to get new or better stuff as we were about our new digs. We acquired about a dozen newer Toyota Hilux and Nissan Patrols to ultimately hand over to them for their fleet to patrol the border.

Why all this, you ask? There was a small but notable terrorist group that was said to be infiltrating Iraq from Turkey. They were noteworthy, but I can't remember the group's name for my life. The point of establishing the IBP was to be a first line of defense for these insurgent fighters and prevent them from getting to the hot spots of Baghdad, Tikrit, Mosul, and the like. I would learn on a later deployment that the hustle and bustle we had come to know in Mosul local life would come to a standstill not due to US forces but to insurgent fighters. Mosul was a busy city with people going about their day-to-day like we weren't even there. The few folks I could talk to on some patrols in the city stated they wanted our presence there to protect them from the bad people.

The beer stores. OK, so we're all human, right? We wanted to have a drink like any average person, so the rules were simple: No drinking when there was work to do. No drunkenness. Dispose of all evidence immediately after drinking. The Kurdish soldiers didn't know what we were up to, which was easy because we all had separate buildings, and none were to enter the others' areas. We stuck with this rotation and didn't drink that much, nor did it become an issue; we were responsible while completing our assigned missions.

For whatever reason, we had about five engineers attached to us, and we let them in on the drinking rules to which they agreed. One night, they were on the roof sharing a bottle of something strong, were drunk as skunks, and decided it would be funny to throw the empty bottle at the Kurdish guys below who were marching by. We informed them that this was unacceptable, and instead of handling it in-house, our newer PSG sent them back down to Mosul. When questioned why we sent them back, they said we were all getting drunk up there in Zakho, and we didn't like what they did on one occasion. They spun a story to make us look like irresponsible drunks who were hating on them or some such BS. It sparked an investigation, and, in the end, the senior NCOs with us got smacked on the wrist and were told that upon return from the deployment, they needed to seek employment elsewhere; most of them went to SF. They took us back to Mosul and sent another platoon to finish ops. No more sweet life of milk and honey. Back to the real war, if you could call it that.

It was November, and the light at the end of the tunnel was getting bigger. We knew that our one-year mark was approaching and that we should receive orders to redeploy home. We were eagerly awaiting that sweet order! Sometime around May, we knew we were in for the long haul, and they started a leave rotation so we could go home for a couple of weeks and get a much-deserved break. I was still on the low end of the totem pole for my discretion, so I wouldn't get to go until November.

I would be home for Thanksgiving, which I was grateful for since I hadn't had any leave in over a year. I was able to see a lot of family that I hadn't seen in nearly three years at this point, but that didn't prevent me from trying to escape the pressure of being around them and running off to the bar with my buddies. I didn't know what I was running from or why; I just knew I wanted to have a few drinks without consequential retribution from leadership.

When I returned to Iraq at the beginning of December, I was thrilled to see that we were laying out gear, getting items cleaned up for customs inspections, and conducting handovers with the leaders of the replacing unit. Their soldiers would be in the country soon, ready to continue the fight. Our last orders of business were to drive all our vehicles and personnel back to Kuwait and have our vehicles inspected by customs, so they could be sent back home. It took three days to travel to Kuwait, but at least there were forward operating bases (FOBs) to stop at along the way that had facilities for eating, showering, and sleeping, as opposed to sleeping on a cot in a tent in the desert the whole way. In Kuwait, customs was extra picky about the details of vehicle cleanliness, as if a bit of desert sand was going to cause a pandemic or something. We made it through that process and hung out, waiting to go home. It sucked waiting like that, but we were able to knock out some redeployment "what to expect" classes in Kuwait before going home and were able to make leave plans while we talked to loved ones as well. I decided to take a whole month of leave to unwind, and since I had all that leave saved up, I was going to use it!

Somewhere amid redeployment prep and leaving for Kuwait, we lost another guy. Initially, he was a problem soldier. He came around throughout the deployment, and with the care of fellow soldiers and teammates, his military career was on the uptick. While out on patrol, their vehicle was struck by one of the first roadside improvised explosive devices (IEDs), or at least one of the first that we had seen in Iraq. With no armor on our Humvees to speak of, he died instantly from the blast. When we returned ten months later, no vehicle was allowed outside a FOB without up-armor, which was only a year too late in most respects. Our fourth loss, and the first one attributed to "combat."

We arrived home in the last week of January to much fanfare, pomp, and circumstance, but I don't remember any of it. All I wanted was to go on leave and get away from all the people I had been surrounded by for the last twelve months. I wanted to put it all behind me, relax with friends and family, and see my beautiful green, mountainous home state of Washington again. I didn't want to see a desert camo uniform (DCU) or BDU for an entire month. I was anxious since we knew we were returning to Iraq in November. We barely had ten months at home and were headed right back to that sand show of shit, so I wanted as much time off as I could muster. I didn't care about whoever wanted to give a speech and congratulate us on a well-done job. No one cared about the mental health classes we had to endure or the other lectures about drinking responsibly and remembering that trash on the side of the road is just trash, not a bomb. We all wanted to see the loved ones that we left behind for a year and not have to worry about the potential of a bullet, rocket, roadside bomb, or any other threat of that nature.

We had family reunions on my mom's side every summer, usually at someone's house on the west side of the state—Seattle suburbs, Ellensburg, and Whidbey Island. Then every few years, those folks would travel over to our side, as my grandparents had a nice spot near us with plenty of space for everyone to hang out. I hadn't been able to take part in this for nearly four years, so during that leave in February 2004, we made the rounds and saw as much family on the west side as possible. It was fun traveling, catching up with loved ones, and seeing the beauty of Washington again. My sister and her family had moved to Western Washington, so seeing how they were doing was good.

I took plenty of time to myself, but my family started to notice a new side of me. It was an aggressive and easily angered side, but only in certain situations, and no one, not even I, knew what would set me off or why. I would snap at whomever for no apparent reason other than the littlest thing having bothered me. My mom asked me to "just go away for a while" a few times as she would hand me her car keys, and I would silently leave the house and go for a drive. All in all, though, leave was good, and outside of those few instances, everyone was happy to see me and glad that I made it home OK.

Returning to Bragg, I was supposed to attend PLDC. It was the first level of leadership training for new or upcoming NCOs, and it was a one-month lockdown joke of a course. Sure, we learned a lot about regulations, leading and taking care of soldiers, drills, and ceremonies, but mainly, we learned how to pass the time while locked down in WWII-style barracks. For most of us, it was a month-long detox and endurance trial in putting up with Army BS. I learned that the dress uniform I received in basic training no longer fit as I was fifteen pounds heavier than when I came in. My pants looked painted onto my legs, and the jacket buttons could pop off if I pushed my chest out. I spent much time in Zakho doing pull-ups, push-ups, and any calisthenics I could. While I wasn't huge, I was in pretty good shape, considering I drank and still smoked off and on around this time. My turn at PLDC would have to wait as my Great Aunt Audrey had passed away, and I wasn't going to miss this funeral even though I had already been home for a month.

I went back for Aunt Audrey's funeral, but this reunion was brief and a lot more somber than the one the month prior. I'm glad that I had gotten to see her in February before she passed away. Even though I knew she was battling cancer, she looked well, and I figured she was winning the fight. It didn't help my emotional state that more than one person said, "We think she was hanging on until she knew you were home and safe." My cousin informed me that Aunt Audrey had this box of prized possessions, and *all* the letters I wrote back to her from Iraq were in it. That broke me. Because I'm not sentimental in the slightest, when we had been getting ready to drive back to Kuwait, I had tossed all non-essential items, including all the letters I'd received, which also included Aunt Audrey's. I felt like shit for a long time because of that.

Returning to Bragg meant resuming the grind. We did some training in preparation for our looming deployment. Many guys put in packets to go SF, flight school, be Ranger instructors, or whatever they could if they wanted to avoid heading back to Iraq. I finally attended PLDC in July and completed it in August, around the same time that one of my teammates got married. His wife returned from California to be with him in North Carolina, even though it would only be a few months before we deployed again. She was a nice, short woman who loved to cook and hated leftovers. She would have him invite me over nearly nightly to have dinner, which I was not opposed to. It beat cooking for myself. She had a co-worker who had just gotten out of a long-term relationship and she'd been telling her about me. She talked me up and said we should meet. Reluctantly, for both of us, we agreed to a double date, which was also a blind date.

"Her name is Genevieve, but she goes by Jenny." My buddy's wife explained. "Her background is Filipino Spanish, I think."

"Does she have an accent?" I asked. It didn't matter to me if she did, I just wanted to know.

"No," she said. "She has a BA in English but doesn't like teaching in a school, so that's why she works at Sylvan Learning Center."

"Is she cute?" I inquired bashfully, knowing women don't care about answering questions like that.

"I think men would say she's attractive." A most political reply.

"OK, I'm in," I said with no reserve. What else did I have going on in life?

We met for dinner at Texas Roadhouse on Skibo Road in Fayetteville, on September 10, 2004. I only remember the day because it was a Friday, and we all mentioned that the next day would be 9/11. We decided to meet early for dinner then shoot pool at Fat Daddy's and get some drinks afterward. Jenny was running late as she went to a college in the area to get some textbooks for her brother, but she got caught in rush-hour traffic on the way back. We had some drinks and appetizers while we waited for her, but the server didn't seem too bothered as the dinner rush hadn't picked up yet.

Jenny arrived an hour and fifteen minutes late. As she sat down, I thought I had seen the most beautiful woman ever, and that was not the rum and Coke talking! I first noticed her incredible smile, which shined with her lip gloss. Her shoulder-length, dark hair framed her face perfectly. Her hair and naturally tan skin complemented her brown eyes, which were light and full of life as she sat down and apologized for her lateness, diving into the reason while we all smiled and nodded. Then, it was as if she was right on time; no one cared, and the conversation flowed. I prepared to order my usual 11 oz Sirloin cooked medium, house salad with ranch, and loaded baked potato. Meanwhile, Jenny ordered first, my exact order except with an 8 oz steak. I was over the moon. She was beautiful and liked steak the same way I did.

After dinner, we went to Fat Daddy's, but our friends took the long route to get there, allowing Jenny and me more time to talk one-on-one.

"You have a newer Mazda 3?" I asked while we set up the pool table, already knowing the answer.

"Yeah!" she replied emphatically.

"Automatic or manual?" I expected her to say automatic.

"Manual," she offered abruptly, with a "no duh" look on her face.

"Oh, nice!"

"I learned to drive a manual because my grandfather had a Dodge Viper that he said I could drive if I knew how to drive a stick. I learned too late, though; he passed away a little after he promised me that."

"I'm sorry to hear that. I'm sure it's a sweet ride."

"I got to ride in it. He sure had fun driving me around a couple of times!" she said joyfully. "What do you drive?"

"A newer RSX. Also a manual."

"That's a nice ride too."

"Thanks," I beamed. "When did you get yours?"

"February fourth, and you?"

"Man, that's crazy, I bought mine the same day." We laughed at the thought of meeting someone who bought a car we both liked on the same day. Serendipitous, kismet.

The conversation shifted from cars to relationships, which neither of us was interested in. We were both side sleepers that couldn't sleep without a pillow between our legs. More coincidences we didn't write off as nothing. After playing a few rounds of pool and only one more beer (I learned my lesson years earlier, right?), we exchanged pleasantries, and I headed home.

I picked up the phone on my way home to call my mom since it was a short five-minute drive to the house.

"Hello?" my mom answered on the other end.

"Mom, I think I just met the woman I will marry someday."

"Oh yeah?" She sounded rather unenthused.

"Yeah. Her name is Genevieve; she goes by Jenny. We had the same steak dinner order, drive manual transmission cars, sleep with a pillow between our legs on our sides. She can play pool just as poorly as me, and she's beautiful too!"

"OK, she sounds great." Again, not sounding as excited as I was.

"Alright, I just wanted to let you know. Talk again soon."

I arrived home beaming at the thought that I had just met the woman who was the perfect fit for me.

CHAPTER FIVE

J enny and I became aware, rather quickly, that our friend-zoning wouldn't last long, and since I was deploying again within two months, we spent every non-working minute together that we could. We'd have lunch together if she had an early afternoon off, and I could sneak away from work. We'd meet for evening dinners and a movie as much as possible. If there wasn't anything else to do, we'd hop over to Barnes and Noble, have a coffee, and peruse some books or magazines. I took her to the range to shoot; she took me bowling, which she was good at but didn't tell me until she beat me three rounds.

I was ecstatic to learn she had a storage unit and asked where I would keep all my stuff. I asked her if she would take care of my prized possessions, including my car, leather jacket, and Smith and Wesson 1911, which I had purchased when we got home. Aside from that, I had a bed, dresser, and stereo that would go into her storage, which I said I would pay for while deployed. Win-win! She would drive my car once a week while I was gone to keep the oils flowing, and everything else would sit where it was.

Jenny had previously been in a relationship with an Army guy. Though he had only deployed to Sinai, Egypt, she was familiar with the leaving process. The morning I was set to leave, I drove my car to her house, where it would remain until I returned. We shared a rather silent car ride to meet up with the rest of the company that was deploying, too. What do you say in that moment? There's not much to say at that point. We both knew what the drive was for. I would be leaving for a year, maybe forever if things didn't go well. The unspoken silence speaks volumes in moments like those.

The company was all abuzz when we pulled up. The hecticness of final preparations broke the uncomfortable silence. Family members were present saying their final goodbyes and though everyone knew what was about to happen, seeing others going through the same feelings in that moment helped break the tension. At least it wasn't a continued silent gathering. Engaging in conversations with family and friends present made the near future seem further away as we chatted and laughed with one another. After all, we were only deploying, not heading down death row.

When the time came for us to depart, the mood was in-different. Sure, Jenny and I were both sad to part, especially after having just met. However, it wasn't the same sadness we saw amongst our friends leaving their wives, regardless of whether they were newlywed or married for some time. It was more like that sadness of leaving a friend for summer break, knowing you'd see them again at the beginning of a new school year. Neither one of us cried, and even Jenny ad-mitted later that she just sat in the car for a minute after we left, realizing that her life would be different than the last couple of months, and then drove home. It wasn't heartless on either part; we just weren't *there* yet in our relationship.

Around mid-November 2004, we were back on a plane to Kuwait. They couldn't wait until after Thanksgiving or give us a Christmas at home. Nope, get ya ass back in the fight. Fine. We had a new TL, and I was the ATL. There were a couple of new guys, and one wasn't cutting the proverbial mustard, so we were trying to get him assigned under HQ or supply so he wouldn't hold us back. In hindsight, our useless TL was the one holding us back. He was a prior Ranger Battalion dude who preferred to kick in doors instead of snooping around and spying. He taught us some good shooting and grappling techniques as he was better than most. Other than that, he was a drunk, the likes of which made me look like a casual drinker. He would grab as many non-alcoholic beers from the chow hall as he could and drink them all as fast as he could to try to get something of a buzz—complete failure.

Once we returned to Kuwait, we couldn't head into Iraq without some up-armor on our vehicles. The accepted practice was to replace the doors with Kydex steel plating, along with bolting steel to the floorboards, seatbacks, and anywhere else bomb shrapnel or bullets could penetrate. We tested the steel plating at the range, and sure enough, it would send bullets of almost any caliber ricocheting off it. We also had steel base gun mounts to protect the gunner from a blast from beneath or a bullet to the face, exposing the rest of him. We only needed to get the vehicles and company to Baghdad or Mosul, I forget which, and we would receive our fully armored Humvees and armored cab LMTVs, which would later save my life. Building suspense and anticipation, you'll have to wait for that story. (Insert smirk here.)

Our team was fortunate and got to ride in a Black Hawk up to the next stop, I'll say Mosul, for argument's sake, to arrange our temporary living accommodations for the company. Once everything was set, we had a couple of days to relax and explore what was in the area before our company's arrival. We were all a bit surprised to find out just how much had been added: MWR tents, phone trailers, PX/Shopette areas for buying necessities and morale items, but most importantly, having access to the phone trailers. We would find great deals on phone cards that would give a good exchange on the minutes and then be able to reload those cards relatively cheap. No more lines waiting for a SAT-phone for the freebie five to ten minutes we would get. Now, if you had the chance to call someone, you could do so without waiting. Internet usage was more prevalent, too, but it was still slow. The United States was beyond dial-up, but Iraq had yet to catch up. Still, it was nice to have easy access to all these options. Kellogg & Brown and Root (KBR) had taken a contract to feed the troops, and the food was the best we had seen. You could eat whatever you wanted, however much you wanted, and since midnight chow had breakfast food, I was all in.

The rest of the company arrived, and we discovered our new destination would be Sinjar. Its location is an hour or so west of Mosul and about an hour and a half from the Syrian border, about midway between the two. Our primary mission would be to stop insurgents smuggling weapons, fighters, and monetary support items like cigarettes across the border. They would smuggle cheap goods across to sell and items to fight or make bombs which would further support the insurgency in Iraq. Before we headed that way, we holed up in a small camp with some SF guys. They mentioned that incoming fire was a significant concern with the insurgency ramping up, but the small area they were in had yet to receive any, so we were optimistic about it. There were daily occurrences of incoming and outgoing fire. The biggest difference was in the sound. The outgoing was loud, as it was nearby, followed by a distant pop. Incoming fire was only loud if you were close. It wasn't a few days, and the randomness of the incoming fire finally tagged our area with the SF guys. There were no deaths, just some shrapnel injuries. It put it all into perspective that the enemy was a greater threat than ever.

We acquired our sweet new armored Humvees and set off to our destination. We had been to Sinjar on the first deployment, but only for a few days. At the time, it was merely an abandoned airstrip with several bunkers for storing aircraft nearby. Other than that, it was desert sand with Sinjar Mountain in the background. Nothingness. This time, things were different. We headed to FOB Sykes, but I didn't have high expectations. We were pleasantly surprised that we would live in containerized housing units (CHUs) in a central area with our teams and the entire company. The chow hall was decent and had all the amenities that the other stops we made along the way had. Comforts go a long way when you still have eleven months ahead. Missions were starting to stack up, and we were ready to work. The first mission was to drive to Mosul and pick up our LMTVs that received armored cabs and suspension upgrades.

It was right after New Year 2005, and we were picking up our rides with some sweet new upgrades. The cab was so solid that the air pressure would change when the doors were opened and closed. The suspension was tight, and even when loaded down, the LMTV felt more poised than it had the year prior without the upgrades. It had a gunner's turret opening, but since we didn't have the armor plating to protect the gunner, we weren't allowed to use it. Fine by us, but we were just a big, enclosed, moving target then. Lastly, they added Recaro racing-style seats with four-point harnesses in case of a vehicle rollover; does that sound familiar?

One of my buddies was supposed to drive, and I would be his TC (truck commander, a fancy term for passenger). My buddy wasn't comfortable driving the LMTV at night with NVGs since he didn't have as much experience as I did, so he asked if I wanted to drive the trip back to Sykes. I was chomping at the bit and jumped at the opportunity. I get it. It's not a fun, fast sports car or anything, but they made it pretty sweet. I wanted the first crack at the bat, so I took it. What else did I have to look forward to aside from being shot at? The infrared (IR) lights didn't illuminate anything, so with a quick call out on the comms, we let the convoy commander know we would have to roll with standard lighting. At least there were only two LMTVs in the convoy, but it was still not tactical. The vehicle drove great. I wished we had those seats on the prior deployment as they were more comfortable, albeit they sat more toward the center, so the driving angle was slightly askew. No bother, I adjusted to that quickly, and we were cruising. Little did we know the cab's dexterity would be tested shortly.

The second vehicle in a dozen-vehicle convoy, we were almost back to Sykes and ready for the night to be over. As we approached what was called Purple Heart Alley, known to be an insurgent hotspot for IEDs and ambushes, I saw the lead vehicle dart to the right of some barricades. As I said, "What the fuck are they doing?" that's when it happened. About sixty feet in front of us, a massive explosion went off. A double stack of 155-millimeter artillery rounds set to explode blew up, signaling an L-shaped ambush. We took small arms fire from the side and the front, all while driving through a dust cloud after being hit by shrapnel from the exploding rounds.

I pressed the pedal to the floor, but the engine had no upgrade to match the suspension or extra weight of the cab. It was slow to pick up speed. I kept the lead vehicle in sight and could see the gunner swiveling to find a target to engage, but no luck. We cleared the dust cloud, I heard a *click* sound, and a chunk of glass went missing from right in front of my face. My buddy had one before him, too, so we agreed it was either sniper fire or some accurate "pray and spray." Thank God for bulletproof glass, or I would *not* be here today. Or maybe parts of my face might not look the same. I'm not sure how many shots we took or what damage was caused by the blasts' shrapnel, but the vehicle's suspension was shaky afterward, and I could see sparks back on the driver's side in the mirror. The passenger side had some flames, and I said, "Shit, the fuel tank is on that side." To calm our nerves, I followed with, "Oh, but it's diesel. It won't explode." I knew I was full of shit, but we both agreed it wouldn't explode and drove on as I prayed, *Please, God, don't let this truck explode!* We made it the last five clicks to Sykes at a rapid speed and barely stopped for the checkpoint, wondering if some insurgents in vehicles were in pursuit; they weren't. After rallying the troops and vehicles, we found that it looked like my truck sustained the most damage, and no one, not one person, received any injuries from the ambush. It did help to set a precedent of vigilance that was not there prior. Major combat operations had ceased a long time ago, but we were still at war against terror on their turf.

My TL no longer wanted to be a team leader and was moved to another position in the company. I was the new TL of Team 3-4. I was hotheaded and thought I had it all figured out. There was no mission too big or too small. I was motivated to prove my worth. I followed some great examples that bucked authority, so I would also show them I was a big boy. Pride comes before the fall; humility is close at hand. Our first mission was up, and we were ready to pounce. It was only three days at most. We were to gain positive identification (PID) on a high-value target (HVT) to facilitate an air strike. We were ecstatic. It was a LRS mission for our team and would provide instant gratification with an air strike. Ready to face this challenge, we forged ahead. Since no one on the team was Air Force qualified to call an air strike, we took two joint tactical air controllers (JTAC). The JTACs assured us they could carry their loads and handle the terrain under low illum conditions with NVGs; I had doubts.

The mission planning went off without a hitch. We had command approval after briefing the finer points and details well. Next up were rehearsals, as we had a couple of days before the mission. The more experienced JTAC was an overweight fellow who looked like he couldn't carry fifty pounds on his back more than a mile, but I gave him the benefit of the doubt. The younger guy looked like he could be a permanent member of one of our teams and carry his ruck and a dead body until his body collapsed. I was usually a good judge of character this way, and we quickly learned I was right. The fat boy couldn't carry his load on flat ground in the dark. How would he make it over some rough terrain in the coming days? Aside from this, the rehearsals went well, but I was still anxious about the movement. We forged ahead with the mission timeline.

We took the platoon and five vehicles to the top of Sinjar Mountain, where a small FOB and a retrans station were located. Their primary purpose was to be an antenna for inter-communication from one side of the mountain to the other. It was mid-January 2005, and it snowed at the 4,000-foot peak. Yes, it snows in hell. I made a snow angel and have the picture to prove it, body armor and all. That night, we would drive halfway down the mountain to the drop-off point, infil approximately eight kilometers on foot, post up for surveillance, and watch for our HVT. Intel stated he was to arrive the following day, so with caffeine pills at hand, we planned to stay up all day. When he arrived, we would conduct the strike, wait until nightfall to exfil and extract, and be done. One shot, one kill, or splatter, whatever a 2,000-pound GBU (Guided Bomb Unit) produces. We were on a high. The glory we missed on the prior deployment was at hand, except for one variable—the fat JTAC.

This fatty couldn't even walk thirty fuckin' meters without slipping, tripping, or falling and making a bunch of noise that set every dog within a three-mile radius on a barking frenzy. We didn't have to worry about being compromised; we never left the gate. All he was carrying was a sleeping bag, food, and water, which proved too much for him. His counterpart had the radios and batteries, the only items with actual weight. We returned to the Humvees, standing by for QRF (Quick Reaction Force) in the event of contact. Informing the platoon leadership of the situation, we loaded up and headed back up the hill to the retrans site to devise a better plan. We looked for an alternate insertion point since that one was blown, but we couldn't see a better location to make it to the SS by daylight. We tossed the idea of not setting up on the hillside and moving to the south side of the OBJ, but the flat ground provided no vantage point. Ultimately, we ran out of time and had to RTB (return to base). The HVT was tracked by his cellphone, and he never went to the suspected meet-up location anyway, so it was all for nothing. Still, it would've been a pretty awesome sight to see the "rockets' red glare" headed to the terrorist-enabling asshole going up in smoke. It was early on in the deployment, so we fine-tuned the things we got wrong and prepared for the next mission.

We weren't at Sykes long until we returned to Mosul to relocate the company to FOB Marez. Marez recently had an insurgent suicide bomber blow up an S-vest inside one of the chow halls. I don't recall how many lost their lives in that event, but they did replace it with a new facility that had even more food options. The gym there was fantastic, and they had a tent set up as a movie theater, which also ran some just-released movies. We all crowded in when we heard they would play *Star Wars: Episode III-Revenge of the Sith* while it was in theaters back home. It was a decent copy of it too. Buying bootleg movies from the local shops was a big thing as we could get some decent new movies, if not still in theaters, for a dollar apiece.

Again, camaraderie was strong as we all shared TVs, Xboxes, DVD players, and anything else we had. I also began working out with some heavy-lifting dudes, and, along with some supplements I was taking, I started to put on some good muscle for the first time in my life. The living conditions weren't as nice as CHUs, but we made it work as we had a large, hard-stand building area for our team. We only slept on cots during our transient periods in and out of country, so aside from being away from home and the constant threat against our lives, the living wasn't too bad.

Care packages came quicker than they had during the previous deployment. Jenny sent me a monthly care package full of whatever goodies I asked her for—usually beef jerky, trail mixes, and whatever else would go well on missions with us in the heat. There was nothing of value for me to send her from Iraq, so I made it a point to send her flowers once a month. I knew that sunflowers were her favorite, so I defaulted to those. However, I would send something a little more appropriate for an occasion, like roses on Valentine's Day.

She surprised me with one incredibly thoughtful care package. She included a scrapbook she had made filled with the few pictures we had taken while together, letters she had written over a few weeks, a phone card with hundreds of minutes to use, and a mixed CD. The CD was loaded with songs that we knew each other liked, and the first song on the mix was Al Green's "Let's Stay Together," which we affectionately dubbed our song. The scrapbook and pictures were nice mementos to return to after dayslong missions in the middle of nowhere.

We did many platoon missions in the desert, trying to determine which villages were friendly and which were hostile or supporting hostile activities. We relied heavily upon our interpreters, as they could read their people better than we could, but we also had to be skeptical of the "terps," because it was possible they weren't accurately relaying information. We had solid, trustworthy terps on every deployment I've been on, so this never proved to be a real concern for us. They had to worry about themselves ,too. If their identities had been discovered, it would have put their families and them in jeopardy. They also wanted to end the fighting and threat to their country and way of life, so most did their jobs phenomenally.

Many of these missions took us to border checkpoints on the Syrian border. We would establish whether the Iraqi commanders at these checkpoints were conducting border security operations satisfactorily, and the terp's ability to read the commander was invaluable. A look around at some of these would reveal the truth rather quickly too. If the Syrian and Iraqi sides seemed to be distant from one another, we figured it was business as usual, and they were maintaining their missions. Other times, we would roll up and see them play soccer with one another across the berm that separated countries and could reasonably assume that under-the-table businesses were being conducted. They may have been friendly with one another to keep the peace. Who knows? I never stepped inside these border facilities as I would rather shoot at the fish in the barrel than be one of the fish myself. I developed a heightened sense of mistrust in Iraq due to the fact we never knew whom we could trust completely.

When border checkpoint and village engagement ops weren't happening, team-level missions sporadically came down the pipe. A month or two later, we had another shot.

There was a road south of Mosul that connected Highway 1 to another parallel road that was becoming the location for more frequent roadside IEDs. We were to surveil the road for five days, and if we saw anyone placing IEDs, we were to call for air strikes by Apache helicopters to neutralize the threat. If we didn't think they would make it in time, we had permission to neutralize the threat ourselves. We decided that handling it ourselves was our only course of action. Wartime glories ahead!

We got to take another Black Hawk flight to recon the location and decided this little outcropping of rocks on a small hillside would be the perfect location for a vantage point for a good mile of the road. There was a FOB nearby with Stryker APCs (armored personnel carrier) that would conduct counter-roadside bomb recons regularly, so the plan was to get dropped off at the FOB, wait until nightfall, roll out with them on one of their ops, walk off the back of the Stryker, and beat feet to our SS in the rocks. No one would be the wiser. We packed the bare minimum food, water, and batteries we'd need, allowing extra space for cold weather gear. It was the end of March, and it was still rather cold at night. Plus, it would rain on us a time or two.

I usually wouldn't sleep for the first two days of a mission. It was always like this. It wasn't for fear, worry, or anything like that. It may have been just that I was excited or something. I would always take this time and let the other guys sleep or rest if that's what they needed, but it usually wasn't an issue for me. On day two, I would always get slap happy, and even just a look could make me, and usually someone else, laugh. That's when it happened.

A small, blue Toyota single-cab pickup truck rolled to a stop on the dirt road about 300 meters before us. I just watched him for a moment, but as a man exited the vehicle, walked to the front, and popped his hood, I thought, *this has to be the signal for a meet-up!*

I alerted the team to wake up and establish comms as we watched and waited for the meeting to go down. We had every weapon trained on the guy and an AT-4 ready in case something bigger came. Claymore clackers were always at hand, two behind us overhead for anyone coming from the top of the hill, one in front for anyone making it past our small arms fire. We were only calling Apaches in the event we needed QRF. That's not what we briefed, but dammit, the day was ours. We were all John Wick-accurate with our weapons, and we were ready to bring the fucking heat.

Within five minutes, a Bongo truck came driving down the road loaded with propane tanks. We were convinced this was it. Adrenaline surged as we eagerly awaited a shootout. It would've been closer to a duck hunt, though, and I envisioned it being over quickly with a couple of well-placed rounds to the propane tanks setting off the lot of them; there had to be fifty of them. The driver of the Bongo truck stopped behind the Toyota, and the two talked for a moment. Anxiousness started rising. *Here it comes, but what will it be? Will they start setting it all up? Will some other guys come out of the back of the Bongo truck? Come on! Get to it! Get to your demise already!*

And then it happened: the driver of the Bongo truck started to drive by the other guy and his vehicle. Disappointment began to kick in. *This guy's truck is really just broken down!* Adrenaline crashed, and suddenly, we all looked like we had just finished a marathon. It was still a comical sight to behold. As the Bongo truck drove off, the other guy, still standing on the dirt road, shook his fist and yelled something at him. It was like watching those old black-and-white Mickey Mouse cartoons with no volume, only movement to signify emotions. As he yelled at him, his truck started to roll backward, very slowly at first but then it picked up a little speed. He turned around to see his truck rolling away, threw his hands in the air in desperation, and shuffled as fast as his apparent old age and man-dress allowed him. We were all chuckling as quietly as we could, all at the slap-happy feeling and seeing it happen similarly, like a silent comedy. He finally got to his truck and got in. It started, and he drove off very slowly. We never saw the truck again. *Bummer; three more days to go.*

On day four, a sheep herder was coming close to our position. A young man and his grandfather, I assumed. That desert will age you quickly; for all I knew, it could have been his dad. He was missing a bunch of teeth, so I think he was older. Once they got too close to our position and knew they wouldn't do anything else since we hadn't seen any nefarious activity, I told one of the guys to get their attention by grabbing them on the shoulder. We were so well hidden that they didn't see us from five feet away. We spooked them so much that they almost fell off the rock in front of us. They would have only dropped six to eight feet, but that was better than meeting a Claymore mine face to face.

We tried to talk with them using hand and arm signals to signify that we were looking for people placing bombs by pointing to our eyes, then to the road, making walking gestures with our fingers, shoveling, and explosion motions. It was pretty genius, I thought. They conveyed that they didn't see anyone doing that by motioning a hard no with both arms. Maybe they thought we were asking if they planted bombs. Either way, nobody saw anything.

We gave them some MRE stuff, and they gave us some soggy flatbread and mushy vegetables, which we didn't eat to avoid getting sick since we had a little over a day to go. They went on their way, and at higher's command, we relocated across the other side of the road at nightfall. In the end, nothing came of this mission either. We were extracted before sunrise on the final day in the same manner we'd been dropped off. Disappointed, we headed back to prep for the next mission. War would elude and relegate us as targets of opportunity for incoming fire. Pity.

Most of this deployment carried on the same way: drives out to the desert for village engagements attempting to establish who the insurgent sympathizers and facilitators were, probes and surveillance near the Syrian border to determine where crossings were taking place and accurately reporting how many and how much stuff was coming through. We would also do impromptu checkpoints on roads near the border attempting to stop the smuggling of men, weapons, and equipment into Iraq. Occasionally, a team-level mission would pop up, but we had our shot at that, and there weren't many more on the docket following. I was OK with this as platoon-sized ops provided more security and were better overall for morale—more people to complain with and share the insufferably hot approaching summer months. As a bonus, if we did come across some nefarious activity out there, they would have thirty-five well-equipped and ready-to-bring destruction dudes to contend with instead of just five or six. We did have some Dohuk trips interspersed within our day-to-day ops, which were spread out well between the platoons. These were good chances to relax and eat some good food, as mentioned before, even though they were usually day trips. One team stayed there for a few months, but that would be their story.

Leave was still a thing, and I waited until May to take my two weeks. I wanted to go home to Washington but also wanted to spend time with Jenny. She wanted to accommodate my leave request and agreed to fly up to see me. We would get to spend time with my family, we would be able to spend her birthday together, and we would all leave a bit happier than we came, albeit I was returning to Iraq. She couldn't get all the time off, so I took her to Manito Park on Spokane's South Hill on May 5th, prior to her birthday and just before she had to head back to North Carolina. Having been in one long-term relationship that didn't survive long distances, I didn't want to miss a chance at this relationship, which could be something great. We had sent each other plenty of letters, and I called her whenever possible. I knew there was no one else in the world for me, and I had serious thoughts about taking the next step in the relationship.

CHAPTER SIX

We have all heard some funny, possibly pathetic, proposal stories that go along the lines of, "So, you want to do this?" It's like deciding to buy a new car or something. My proposal to Jenny wasn't that sad, but it wasn't as elaborate as some promposals that kids do these days. After we walked around Manito Park on Spokane's South Hill for a while, taking in the sights of the gorgeous tulip, rose, and other gardens they have there, I strategically guided us to a more secluded area where I had planned to propose. Knowing the moment was close, I began to lead the conversation in a direction that could hint something more was coming.

"So, it's been great getting to know more about you since we've been together, despite the long distance." I opened, unsure how to handle the situation since I had never proposed to anyone before.

"Yeah," Jenny replied. "It's had its challenges, but we've managed to keep in touch and grow our relationship." She smiled reassuringly.

Does she know what I'm leading to? I pondered her response as I continued to lead her to a pathway I saw on the side of the large rose garden, where no one was walking.

"I think we hit it off! I feel more deeply about you than other relationships I have been in," I said struggling for words. I was saying anything that felt right or at least felt like it was progressing the conversation.

"I feel the same way. It's easy to be with you," she said, my confidence building because she felt the same.

After a little more awkward, simple conversation about our relationship, feelings, and finally finding a location away from people I decided it was time.

I turned to her and stated matter-of-factly, "I have a surprise for you, but you have to close your eyes."

"Uh . . . OK." She hesitated but closed her eyes anyway.

"I love you and can't imagine spending the rest of my life with anyone else." I exhaled. "OK, open your eyes."

I was there before her with an open box and a beautiful ring. When she saw it and realized what was going on, I said, "Will you marry me?"

She paused for a moment, later explaining she'd been stunned, and shortly after said, "Yes!"

I slipped the ring on her finger, which fit a little loose but was glimmering on the overcast day, and she proudly wore it with an equally glimmering smile for the rest of the day. We walked around Manito Park for a while longer, posing for photos that we asked other people to take of us to commemorate the moment in history forever.

We decided that we should have a date set before I headed back to Iraq, so after much deliberation about seasons and locations, we settled on July 8, 2006. We figured we would find something in Wilmington, since we both liked that area.

Having recently proposed, setting a date, and knowing that in a little over a year I would be getting married to a woman whom I loved and found amazing, I went back to Iraq a rather happy guy, even though I still had about six months before I would make it back to my beautiful fiancée.

CHAPTER SEVEN

I returned to Iraq enthralled and excited over my pending nuptials and fully immersed in the challenges of helping to plan the big day while being thousands of miles away and continuing with missions. Jenny and I had discussed the details of the date, time, location, invitation list, and all the usual stuff before I left for my deployment. While in Iraq, Jenny and I were on the phone talking about having children, when I asked jokingly, "How old do you think a kid should be before you hit them with a closed fist?"

Silence.

"Hello . . . babe? Hello!" *Did she not understand I was joking?*

I had used the phone trailer so much that I had memorized every number I needed to enter to make a call; I frantically redialed to get ahold of her to make sure she knew I was joking. It took seven redials before a reconnection, and she answered with, "Hey," to which I cut her off immediately, saying, "You know I was joking, right?" She said of course, but we laughed hysterically at the situation and knew it would be a story for all time.

While the remainder of the deployment didn't change, I was a little more lighthearted for the rest of it. The mission sets were the same, but there was some peace in consistency. We knew what we were doing and when we were doing it. We were told that all our intel on the border happenings was getting to the President's desk so he could make informed decisions on applying political pressure on Syria to stop supporting the insurgency, which gave purpose to our mission and felt pretty good. Whether it was true or a lie sold to us to motivate us to continue the mission didn't matter. I took it as a truth, and again, I enjoyed the brotherhood forged in the fire of combat with these fine gents.

I attended another promotion board to make staff sergeant (E-6) and did well on that board even though I couldn't hear the representatives most of the time. It was June, in an open bay with no AC. The garbage ceiling fans set to the highest speed drowned out almost everything everyone said. I had already lost significant hearing in my left ear, and my right had been trying to follow suit for a while too. There was no feasible cause for the hearing loss as it baffled every ear, nose, and throat (ENT) doc I had seen. It happened sometime in 2003, while in Iraq. I was leaning over to get something out of my rucksack, and after I stood up, there was a loud tone, like an unending, high-pitch beep that had never gone away since. It was also as if the volume had been permanently turned down.

The board members probably got tired of me asking them to repeat themselves, which helped end it quicker.

"How many levels of mission-oriented protective posture gear are there?" one board member asked, somewhat muffled by the fans.

"Five," I replied confidently.

"WRONG!" he blasted back, trying to rattle me. It didn't work since I didn't care what I got right.

"What Army regulation covers counseling?" he continued.

"AR 600-1."

"WRONG! He doesn't know anything!" he blasted again to a smirk on my face.

The board went on and on with standard questions, and then the SGM asked his. He was a quiet talker, and I could not hear anything he said. I knew he covered current events, so I tried to guess what he said.

Quiet, muffled questioning, all I heard was, "Michael Jackson?"

"Uh, I'm not sure, SGM, but he's had a successful music career, so whatever is going on can't be that bad." He stared dead at me, and the rest of the board members chuckled lightly and shook their heads. *What did Mikey do this time?* I wondered.

Again, more inaudible words. I couldn't even pick out something to base an answer on.

"SGM, it's hard to hear you over the fans. Can you speak up, please?" You'd have thought I asked the pope to drop a baby instead of kiss one. Every board member gave me a "how dare you" stare—just because the guy outranked me by a lot.

He spoke slightly louder, and I heard, "Storms in the South."

Remembering hearing something about that on a news channel during dinner, I said, "Well, if it's severe enough, the people in the area should evacuate. However, if they have lived there their entire lives, they may know the area well and how it will affect them. They're adults who can make and be accountable for their own decisions." I felt proud about my answer, but the board members stared blankly at me again. They must have wanted me to have a definitive answer like, "Make them evacuate now!"

"OK, Sergeant Kruger, that's enough for now. You'll have the results within a week. Dismissed."

After the SGM finished, I stood, saluted, and turned to walk out the door.

I passed the board just fine, but I didn't get promoted. I had planned to do my six-year enlistment and move on to *anything* else. I didn't have the one year remaining I needed to get promoted—I would have had to extend for two months or reenlist to get promoted. Since I had no desire to return to Iraq, I told them they could keep the rank even though a decent, tax-free bonus came with reenlisting.

We were about to head to Al-Asad in mid-September to start switching out with the incoming unit. We were headed south to occupy space on Al-Asad airbase with the Marines and begin our left-seat, right-seat rides with the incoming National Guard LRS unit that would take over ops in the AO. I was excited for several reasons. One was seeing the Euphrates River, which was no more or less impressive than the Tigris, which we had seen multiple times. The airbase had a pool, and everyone was thrilled about that since missions were to be for the transition of responsibility, and we would have the time to use it. Any norm change was a cause for celebration, so I enthusiastically accepted moving with the entire company there as it was one step closer to going home.

Living with Marines was a new experience though. They had all sorts of extra rules in their area of the base that we hadn't experienced anywhere else.

Driving into the Marines area, you had to have a ground guide for your vehicle, no exceptions. (A ground guide walks in front of the vehicle and motions you to stop in case of an obstacle.) If you conducted outdoor PT, they required you to wear a PT belt. Yes, in combat. Sleeves had to be unrolled and cuffed while in the DFAC. They had two poor privates correcting everyone who didn't meet the uniform standard. Seriously, I can't make that one up!

Water usage was restricted to so-called Navy showers: use enough water to get wet, then turn it off while soaping up. Turn it back on to rinse off, done. Water usage for shaving was similarly enforced. Missing saluting someone because you weren't paying attention forfeits your life. It was the equivalent of going from living in a house with rather lax rules to moving in with a friend you don't like that much and who has super-strict parents. Keep your life-style to yourself, and we'll keep ours.

On one of our final missions out there, we encountered a HVT whom we were to detain. The HVT was allegedly so high value that we were to treat him almost like he was royalty. Well, desert royalty at least. We gave him a cot to sleep on, water, and food, and we didn't even restrain him with zip ties. We were told to stay in position, and the Marines would send a helicopter with some troops to pick him up. It turned into quite the shit show as the helicopter came at night while it was low illum. I watched them try to land, but there was so much dust that they sat at a low hover for a few minutes. Hovering up and down ten feet for all that time, they finally decided to plant it, which broke the skids out from underneath the Huey they were flying. *Don't worry; we have cots and food for you, too. Oh, and we'll pull security as well. Thanks, Marines.* We passed word to higher and they informed us that a wrecking crew would be out in the morning.

A little after dawn, three Marine assault amphibious vehicles pulled up and dropped the rear ramps. Dozens of Marines came *hut-hut-hut-ing* out the back in full combat gear, weapons ready like they were about to storm Nor-mandy, not as if they arrived at a secure location. They set up a perimeter inside ours and around the helicopter. I was next to our PSG; we looked at each other while we all had our ball caps on, no body armor or helmets, M-4s slung and dangling. I asked him if he wanted me to pass the word to get ready to move out. He snickered and said, "Yes, please," to which I quickly obliged, and we handed everything over to the militant Marines.

We were supposed to have another mission after that, but it went to another platoon. We were free and clear to wait until the trip to Kuwait and then home. We all hit the gym, the pool, and the DFAC and watched movies and TV shows to pass the next couple of weeks a bit quicker. As with the previous deployment, there was a bit of excitement combined with anxiety before going home. I was glad to know I'd be home for Thanksgiving, which I would celebrate with Jenny's family, and then to Washington for Christmas. Jenny would accompany me as my fiancée, and there would be no BS training interruptions in between. We would finalize the wedding plans, decide where to live after that, work on my next career post-military, and everything. Everyone would be happy and live happily ever after. *Roll credits. Goodnight, folks.* And then there was reality.

Redeployment is no joyride. Everybody builds up a dream scenario of the foods, places to go, things they'll do on leave, etc. But these expectations may end up being larger than life. Soldiers coming home want to regain the freedoms they lost for a year. Spouses and other family members' expectations may be different. They're happy to have their spouse back, but you don't spend a year away in a different environment and not change at least a little. Family members notice this more than the soldiers do. Plus, everyone wants time with the soldier coming home and most aren't willing to wait.

My mom really wanted to be there when I got home this time, as she had missed my return on the previous deployment. She and Jenny had been working together to get me an apartment lined up, but this fell through because I would no longer be receiving the Basic Allowance for Housing (BAH) that I had been getting since the prior deployment. Since there was room in the barracks, only SSGs and higher ranks would receive it. My non-promotion plan was already backfiring on me.

Adding to the stress of coming home, Jenny had to pick up my mom from the airport, and her flight was delayed.

This caused both Jenny and my mom to miss me coming off the plane and the initial hugs, kisses, and greetings we all get to do after a short speech from someone who didn't deploy and thinks we care what they have to say. We don't. We just want to turn in our weapons and go home with our families. Jenny was mad because she missed out on that. I was frustrated not because she wasn't there, but because she couldn't let go of her anger during the ordeal.

Continuing in frustrations, I didn't want to stay in the barracks, and I didn't have an apartment, so I stayed in the hotel with my mom. I didn't want to do that, but I didn't want to leave her there by herself either. I got my car back from Jenny, so I was the only ride my mom would have too. My mom got to see me come home from a deployment, even though I would take leave in a couple of months to go to Washington for Christmas, but I couldn't spend time with my fiancée since I had to make sure my mom wasn't floating in the breeze.

After my mom left, I stayed in the barracks and made the best of it since it was only for a bit longer. Being back and wanting the freedom to breathe and relax was met with the strain of needing to plan for those long-term things mentioned, and I didn't want to deal with those just yet.

"We want to get married in Wilmington, but where?" Jenny started.

"I don't know," I said, not wanting to talk about it.

"OK, indoor or outdoor?" she continued, looking for some answer to get the ball rolling.

"Indoor!" I exclaimed. "It'll be too damn hot outside in July." I calmed a bit, sensing my tension rising.

"Indoor. So, the beach is out."

"I guess so." My dismissive reply.

"What type of venue do you want?"

"I don't know." More dismissal, hoping the subject would change. I knew we needed to plan the wedding, but I had only been home for a few weeks.

"We need to start planning this, or we may not have a lo-cation, caterer, photographer, or something!" Her tone was rising, near yelling.

"We will. I just want to ease into it . . ."

"We don't have time to ease into it!" She cut me off.

"Fine, let's figure it all out now!" my tone matching hers.

"Forget it; we'll just do it all when *you're* ready," she snapped sarcastically on her way out the door.

Incidents like these became more commonplace as I wasn't as focused on the wedding as she would have wanted me to be. I still had work requirements, so my mind was split on what was more important. I also had the stress of getting out of the Army around the time we were to be married, and Jenny didn't want to continue teaching because she didn't enjoy it and North Carolina paid teachers poorly. I would cope by drinking, which ramped up because of the pressure I was starting to feel. After one too many unacceptable answers, along with more drunkenness and selfishness on my part, Jenny called off the wedding a few days after New Year 2006.

CHAPTER EIGHT

A week after New Year, Jenny and I were done. A buddy was having a birthday party that most of the company was going to and I decided it would be an excellent time to drink my blues away and get my mind right. Oxymorons, right?

I got so drunk at the party, yet I never blacked out. I wish I had so I wouldn't have the memories of what an ass I made of myself. As usual, we started by drinking, hanging out, and enjoying good food and music.

"KRUG! What's up, man? How are you tonight?" someone asked as I walked in the door.

"Well, I'm alone if that's any indication of an answer," I shot back, seething with sarcasm.

"Oh, me too! Though it looks like I'm headed for a divorce."

"Sorry to hear it." I lied. I didn't care about their situation, just mine. "I brought a case of beer to help us all forget about that!"

"Put it in the pile with the rest!" someone else said. Three other cases were in the kitchen next to liquor and wine bottles.

"KRUG! Pour me one," or "KRUG, toss me a beer," I heard as I added my stock to the pile.

"Where's Jenny?" another asked.

"Not here, and I don't think we will be a thing anymore," I said, desperately wanting to change the subject.

"Oh, that's too bad." An insincere response.

"Yeah," I replied, not sure of anything else to say.

After some barbecue and plenty of alcohol to help me forget my relationship issues, we ended up going to The Palomino. It's some dump of a country bar in Fayetteville. My first time, also my last. I was never into bars, strip clubs, or nightlife. Not to mention, I almost got tossed out of The Palomino while trying to ride the fake horse behind the fence. I kept trying to snatch beers out of my buddy's hands with some fast motion like trying to disarm an assailant, and in general, just being a drunken jackass.

Finally, after we got back to my buddy's house, I was taking a piss off the back door stoop and thought it would be funny to start spinning around and pissing inside the door, too. My infantry buddies laughed hysterically. His girlfriend's friends thought it was too much and left. I begged for my keys so I could go home. At this time, I had enough of the barracks and got that apartment anyway, but my buddy wouldn't give me the keys and made me sleep it off—probably the best decision of the night.

I can't believe what I did. I could blame alcohol, but I was the one who wasn't in control. I didn't control my drinking; I allowed my emotions to get the best of me, drive me into drunkenness, and act like the biggest idiot I've been, if not ever, at least in a long time. I embarrassed myself in front of most of the company, and all I could do was hope that they were all drunk enough to think it was funny too. I didn't want to be that guy anymore. The one who was the drunk that everyone looked at to see what crazy, stupid shit he was going to do and have stories about. Sure, at some point they're funny, both the drunk and the stories. But then it becomes pathetic. Like, why can't that guy get his life together?

As I drove home the following morning, I knew this had to change. *I can't keep going like this,* I thought as I turned off the radio, driving home in a hungover stupor. *I love Jenny and want to be with her.* Right? Yes, of course I do! *We've had some great times, even while being apart. We were faithful to each other, communicated as often as possible, and even grew in our relationship over twelve months apart. I don't want to throw that away. I don't want to start over with someone else.* Right? No! *Well, what are you going to do?* The final thought I had as I arrived at the apartment and stumbled up the stairs and directly to the shower.

I quit drinking cold turkey, committed to winning Jenny back, and overhauling my life. All I had to do now was make it happen.

CHAPTER NINE

In February, even though it looked like I wasn't getting married, I decided to re-enlist. One of the main things Jenny and I argued about before breaking up was that I shouldn't get out of the military because I had no real plan, only a backup plan. If nothing worked out, I would work with the Border Patrol since I was a prime candidate for that. *Military it is,* I thought as I looked at the re-enlisting bonus. This time, it was even higher than the bonus offered had I re-enlisted during the deployment. Even though it was taxed, I made more than if I had re-enlisted previously. I was able to pay off my car and also used some to get furniture for the apartment. While re-enlisting wasn't something I truly wanted to do, I knew that Jenny wanted income stability, and this was a way to provide it. I was becoming a real boy!

After re-enlisting, I went to the promotion board again and made E-6 in March or April. The company needed more HALO JMs, so I went with our PSG, Blizard, who we called "Bliz," in April and had a blast in Yuma. He liked being there with me since he had a designated driver and would hit the Veteran of Foreign Wars (VFW) hall or local dive bars, and I would drive him back safely.

"You know Krug, there's an old saying," Bliz started and then paused to collect his thoughts, as he'd had one too many. "Never trust someone who doesn't drink." He finally finished his thought.

"Is that right?" I replied. Knowing he was drunk, and the shit-talking was going to commence.

"Yep!" he blurted. "Why don't you drink? Can't handle your alcohol?"

"Nah, man, I can handle it too well until I can't."

"Is that two or three shots?" he snapped back with a laugh.

We had just arrived back to the hotel and were walking toward the door to head upstairs to our rooms.

"Two or three shots isn't even a good pre-game warm-up!" I shot back, wondering why I was even defending myself. By now, we were making our way up the stairs.

"Oh, a six-pack then. Is that all you have in you?"

"Look, old man, you've barely had more than a six-pack yourself tonight, and there's no way that should be causing you to stumble up the stairs. Keep talking shit and I'll throw your old ass down the stairs after we get to the top and tell everyone you fell. Shit, you probably won't remember anyway!" I finished with a loud laugh.

"Krug," he said, long and drawn out like it had a *dozen u's* in it, "don't throw me down the stairs!"

I laughed again as we made it to our rooms. "Sleep well, old bastard."

"G-night, Krug," he drunkenly muttered.

He didn't remember the conversation the next morning but laughed about it when I told him. "Sounds like something I would say, I'm glad I didn't wake up in the hospital after 'falling' down a flight of stairs!" he stated with finger quotes.

Jenny and I hadn't stopped talking outright, but we weren't getting together for anything either. I had started going to Manna Church in Fayetteville with Jenny when we were still together, and kept up on that, though I would go at a different time than the one I knew she went to so she wouldn't have to see me. I had prayed about what I could do to win her back, and I received an answer as plainly as I heard someone talking to me in front of my face. Remember the scrapbook she sent me when I was deployed? I knew I had to make one for her. I called my sister and asked her how to make one since she also took all my photos from my deployments and put a book together at my request. After getting some details of what I wanted to create, she sent me to Michael's and Hobby Lobby for what I needed, and I was off! The scrapbook came together quickly and easily. Jenny and I love Scrabble, so I used Scrabble letters to label pages. I typed and printed out some of the letters we had sent each other and made a mixed CD of songs that reminded me of her. It was extraordinary for my first attempt at this sort of thing, and I planned to give it to her on her birthday, May 10th.

Our previously planned wedding would have been two months away. We had canceled all the vendors and had only lost our minimal deposits. I wasn't trying to ruin her birthday, but I wanted to give this to her and show her that I cared deeply enough to make something special for her rather than some meaningless, store-bought flowers and a sappy card. I was sober and had no desire to drink. I said a quick prayer, asking for a sign. *If she dismisses the scrapbook as a nice present with a "thank you," I won't bother with any more wedding talk. I'll understand it's time to move on. If she loves it and maybe tears up, I'll re-engage all the wedding stuff.*

Well, guess what? She broke down in tears. She loved it, every last page of it! After some non-wedding talk about our relationship, I proposed that we continue with the wedding plans. I hadn't told my family it was off yet, so they all planned to come. It wasn't because I thought we would be back together and on the same timeline; I was ashamed of myself and never bothered to mention it.

None of the vendors had removed us from their schedules, so we were still a go for July 8, 2006, at the Saint Thomas Preservation Hall in Wilmington, North Carolina. If that wasn't by God's hand, then I don't know what is.

Of course, nothing comes easy, even if it is by God's hand. I had an opportunity to go to Pathfinder School at Fort Campbell, Kentucky, and about a dozen of us jumped on that. Most of the wedding plans were complete, so that wasn't an issue. Jenny was anxious about things not going to plan and she felt overwhelmed by the what-ifs that kept popping up. "What if family is late showing up?" "What if the dress alterations aren't done on time?" "What if a vendor is missing something?" Or anything else that would cause a cataclysmic domino effect and destroy the happiest day of our lives. Me being hundreds of miles away again didn't help the situation at all. I would usually have some late calls, long calls, or both to help calm uneasy nerves about the days ahead. I would return from Fort Campbell with about two weeks left before the wedding, and everything seemed to be coming along just fine.

Wedding finalizations aside, Pathfinder was a mental smoker. The Air Assault course, also at Campbell, is referred to as "the toughest ten days in the Army." I learned some years later that no, it is not. Pathfinder is the equivalent of a graduated Air Assault course. On day one, they gave us a twenty-five-page manual containing information to memorize for the first test the following day. The day's classes were based on the material in the manual, so we had to review it that night for the next day's test. With the first test down, it was on to more of the same, but the manual was three times the size and would be broken down into the next two days—a test on each day. Fail one, you get a retest; fail that, see ya later. All the material was on rigging military items for air transportability.

We had to determine which strings, ropes, or straps would be strong enough to support various types of loads, including equipment, vehicles, and odd-shaped items such as artillery pieces. We had to learn the density and strength of the various ropes and cords or whether chains were more appropriate for the rigging. Finally, it culminated in inspecting different loads with improper rigging, whether due to the wrong type of material used to rig the load or improper rigging at some point. After passing all these tests in the first week, we would learn how to set up a LZ and DZ for Air Assault and airborne operations.

Setting up LZs and DZs was much more straightforward than inspecting loads within a given time, which reminded me of the JM course back at Bragg. DZs were simple; the type of aircraft and the number of jumpers determined how big a DZ needed to be. The winds determined the exit point. Done. Helicopter LZs were a bit trickier. Chinooks needed more landing area than a Black Hawk. If you had both aircraft landing, you needed sufficient room for them without causing an aviation mishap. Were they landing loaded and taking off empty? They would need to land at the far end of the HLZ to give them extra time for all the weight. They required more takeoff room if they were picking up troops or gear. There was no official measurement for one meter except for the elongated stride of an average-height person. Take a long step, and there's one meter. Easy enough.

After passing all these tests, they had a final test that covered all the material before the field day, which consisted of breaking the class up into two groups and having them take turns as a primary leader or alternate leader and setting up a helicopter landing or pickup zone in a specified amount of time. I think it was ninety minutes, but I can't be sure; maybe it was only an hour. Two of my buddies and I were in the running for honor graduate until the final. As we returned to our rooms, they wanted to get together to study for the final. I said, "If we don't know it by now, we won't know it tomorrow. Let's go to the pool on post and have some fun!" They laughed and agreed, and the next morning, we all failed the final but passed the retest. They gave me shit for that for a while, but I only presented an option; they were the ones that took it. Anyway, with honor grad in the rear-view mirror, we prepped for the field day.

No one wanted to fail the field day as it was literally the last day. You would have to redo the entire course if you failed. We started the course with forty-four people and were already down to twenty-two. They broke us up into two groups with two instructors. We would set up HLZs until all students were complete. It was mid-June in Kentucky, hot and humid.

We considered ourselves fortunate as the two instructors we had with us seemed to be the laxest of them all. Some instructors were "protectors of the badge," meaning they prided themselves on failing people and would look for any minor infraction to send them packing. I had a run-in with one of them during the inspection of a HHMWV mounted for a sling load. I knocked one of the transmission cover lockdowns out of place and annotated it as a deficiency. Neither one of us noticed whether I caused it or not. He swore that it wasn't a deficiency and that I failed the inspection. Once the lead instructor was brought over to mediate, at the other instructor's request, he stated that it was a deficiency and that I annotated it, so I was good to go. Phew, I lucked out on that one. We didn't have that guy for the field, so I felt better about all our chances.

One guy in our group failed his first chance as team lead but got another one since we had eleven guys. That meant someone would have to go twice anyway. He passed the second time, and we were driving on with the task at hand. It was now my turn.

I had to set up a pickup zone, which meant the landing areas had to be at the close end of the open area to give the helicopters extra room for the excess weight to take off. The HLZ had to accommodate four Black Hawks and one Chinook. Even though it was three o'clock in the morning, I was ready for the setup. I set the guys in motion and began my steps to emplace the markings for the HLZ. We knocked it out with plenty of time, as we had refined our technique throughout the day, and the instructors were ready to grade. As they started to measure everything out and checked the setup, I could see nods of approval in the flashlight beams, and then, after a little while, they stopped dead in their tracks. I saw them talking, and my heart started racing. *What did I do wrong?* I thought. *I don't want to do this again. Ugh, we all just want this day to end.* As if the exhaustion wasn't enough, the chigger bites we all had on our legs were starting to annoy the mess out of us. They finished up with their measurements and returned to the group.

One of the instructors asked, "Was this supposed to be a LZ or a PZ (pick-up zone)?"

My heart sank. "A PZ," I said. *Dammit!* I set up an LZ. *Round two, here I come.*

He grabbed the paper, crumpled it, got a new one, wrote LZ on it, and said, "Good job, you pass. Let's get the next one up because I want to go home, too!"

Only one guy in the other group failed the field day— some infantry major. Of course, the "badge protector" got him. I don't recall what happened as I only heard the story second-hand, but I remember it sounded like BS to all of us.

Pathfinder completed; we headed home. It was time to get married!

Fast forward to July 8, 2006, and you find me setting up tables and chairs at my own wedding reception.

"Are you the best man?" the DJ asked.

"No, I'm the groom," I replied.

"Why are you setting those up?"

Good question.

Family will let you down sometimes, and some of our family who said they would come early to help set up informed us the night before that they were on vacation, and no one was going to dictate what they did with their time. *Thanks for the backhanded offer to help.* I kept my frustration in check to calm my bride's nerves and informed her that I would set up for the ceremony while she was getting dolled up. The caterers said they would set up the reception, which would only take them an additional thirty minutes or so; no extra charge. Crisis averted, mental notes of lying family members made, nothing was stopping this wedding from happening.

It was a small wedding with family and a few friends. We had a typical church ceremony, followed by intermissions with hors d' oeuvres and sparkling lemonade drinks while the caterers set up.

About this time, the photographer pulled us aside and said, "Hey, I don't want to ruin any moments, but I wanted to make sure I got the check before the festivities began, as it tends to be forgotten."

"No problem, we'll be right back."

My wife and I went down to where all our bags were, and just being caught up in the moment of everything while I was writing out the check, I asked, "What day is it?"

Jenny looked like she was about to turn Bridezilla for a moment and said very sternly, "I don't know, *what* day *is* it?"

I laughed it off, realizing how stupid the question was, and said, "Oh yeah," sheepishly. She got over it quickly, and we were back to the celebration. I received some re-demption years later when we shared our story with some friends, and she mentioned that July 10 was our wedding day. Laughing, I blasted out, "*What* day *was* it!" Now we get to give each other crap for it. Off the hook for that one!

We had a nice spread of food and a chocolate fountain—a must-have for Jenny. It was perfect as we got plenty of pic-tures with our nieces and nephews with chocolate-smeared faces. All the wedding staples were present: cake cutting, arms locked, Welch's grape juice drink, tossing the bouquet and garter, first dances, and then the DJ ran the show for the next hour or so.

After the reception, we went around downtown Wilming-ton to get pictures in various locations with the photographer and had a great time. When we returned to the church, I was surprised to see my car covered in most of the decorations from the church, "JUST MARRIED" across the rear window and hearts drawn in washable paint on the windows and tail-lights. As we exited and rice was thrown, Jenny's uncle threw a small, unopened bag of rice at me and hit me square in the forehead. I was so distracted by all the rice flying that I didn't realize it until he told me later, and we laughed about it—I'll try that one myself at the next wedding I go to.

We had our honeymoon in Wilmington and spent most of the time at the beach, though we did explore the area a bit up and down the coast too. It was a great time, to be sure, but as the saying goes, all good things come to an end, and life was about to throw countless tests at us. Some of them broke us completely.

CHAPTER TEN

Immediately following our wedding, I had to attend Basic Non-Commissioned Officers Course (BNCOC)—one of those Army wastes of time they have you attend to ensure they are training you to be successful in the next rank. At least I was able to do that at Fort Bragg so I could be home every night with my wife. Shortly after this course, over a dozen of us attended Reconnaissance and Surveillance Leaders Course (RSLC). Truth be told, it would have been a better course to attend when I was less experienced in LRS but it was still good training. Since we had a good chunk of the junior enlisted in the course, I was able to help them grow in their understanding of how we work. Bonus, we were able to static line jump both a C-130 ramp blast and hop out the side of a Black Hawk for a jump while we were there, something I had never done.

But the good times were about to end because we knew we were headed back to Iraq, and while I'd been home nearly two years now—something Jenny and I were so grateful for—a twelve-month deployment was looming on the horizon. I was growing tired of the infantry game, though LRS life had been good compared to regular line infantry life. Through this, I was ready to start looking at alternatives, as deployment wasn't my favorite idea of life as a married man. I decided to work on my warrant officer packet for flight school, which was somewhat short-lived at this time.

The first order of business was to pass the Alternate Flight Aptitude Selection Test (AFAST), which tested your knowledge and ability to quickly assess mechanical situations and patterns of things in various forms. You had to score above 90 to be accepted into flight school, with a maximum score of 130. I got a 124. I was good to go.

"Hey Bliz! I just passed the AFAST with a high score. I'm going to work on my warrant officer packet for flight!" I said with great enthusiasm, expecting to be met with the same response.

"What?" he replied quizzically, almost angrily.

"I want to fly helicopters. Kiowas specifically, but yeah, I passed the test and am going to put together a packet to submit to be a warrant officer."

"I'll tell you what, Krug, you do that, and I'll send you to S3 (operations shop) and you can work for Jim and schedule training and jumps until you get picked up for warrant. You won't have a team, and you'll sit at a desk until you leave."

I stood there, motionless, silent. I thought we were buddies even though he was my PSG. *Why would you say that? Aren't you happy that I'm moving on, moving up?* I wish I had asked but I stood there like a beaten puppy waiting for the owner to hand him a treat and pat him on the head.

I attempted to formulate a response, but Bliz cut me off, "If you stay and give us one more deployment, you'll be the go-to team for the duration, get whatever missions you want, and will be able to call the shots. I'll make sure you don't get messed with."

This didn't really appeal to me at all, but I didn't know what else to do. I felt pressure since now I knew people were relying on me, more than just Bliz. I had a team that I built and it was a group of great guys I didn't want to abandon to someone else. I was proud of all the work and team building we had done through blood, sweat, and possibly some tears. I didn't want that to go to waste either. Reluctantly, I obliged his passive-aggressive request to stay.

Now that I think about it, I reluctantly obliged much of my military career, which is possibly why I had so much bitterness in the end. Hindsight, right?

New Year 2007, we knew we were deploying in six short months near the end of June. We had some team training coming up, but nothing that most of us weren't already accustomed to. It was mainly to get the newer guys some experience, and since the company shuffled team members around to distribute the fuckin' new guys (FNGs) evenly, we all needed field time to reestablish cohesion and teach them the ropes. I had a team I could trust, with two solid E-5s and two more ready for the promotion board that would have to wait until the first months of the looming deployment. I wouldn't have traded anyone on my team as we operated as one sound mind to accomplish the mission and even knew what each other was thinking. It was the closest I had ever been with five other individuals and probably the closest I would ever be to a small team since then.

Getting ready for deployment, at least for me, also meant getting PRK laser eye surgery. The recovery was four days at home, and I had to take pain meds and eye drops, doctor's orders. If you haven't had this surgery, then I can inform you that even the toughest guy would need every single one of the pills and drops. I didn't need to be led around the house like a blind man or require help to go to the bathroom as some people have said they needed, but every time I opened my eyes for the first time in the morning, it felt like someone was holding a lighter up to my eyeballs. That lasted for the better part of a month, and I could not do any training anyway.

I was glad to have the surgery done when I did, as my ATL, who was ready to be a TL, had no problems stepping

up. I felt a sense of pride knowing that I had trained someone to be good at their position while being ready for the
next one should the need arise. Most of my training focused
on being good at my current position rather than receiving
solid mentoring for the next one. I usually held the next
position before I had the rank, so my learning curve was
always trial by fire. I'm glad I could save a few guys from
that experience.

The closer the deployment got, the more information we
were given, to include this lovely update by the battalion
rep: "For those of you who aren't aware, there is a surge to
send troops to Iraq to combat the massive insurgent threat.
This means that large numbers of troops will be deployed
to Iraq for extended periods. We wanted to inform you that
the deployment is slated for fifteen months at this time."

What the fuck! I leaned back in my chair. *Fifteen months?
I knew I should have just gone with the warrant officer option,
damn it!*

Turning to each other with shrugs and discontented eye
rolls, the battalion rep continued.

"There is something good to take from the order, though.
It states that it is not less than twelve months, but not longer
than fifteen, so it may only be twelve months."

Some more muffled laughter and, "Yeah, right!" could
be heard. I snickered, having flashbacks of 2003 when we
heard every month that we were *probably* going home the
following month for three months straight. I left any hope
of fifteen months turning into twelve at the door.

"Finally, everyone here today is going forward to the
sandbox. We have been fenced in since the new year, and
any packets to SF, schools, or anything else will be kicked
back." The rep finished up and asked for questions which
was met with more muffled murmuring amongst the crowd.

After leaving for the day and arriving home, I had to tell Jenny the news. She knew of the deployment, but not about the updated timeline. I took a deep breath and began, "Hey babe, they have the deployment dates, finally."

"Finally! When are you leaving?"

"The end of June."

"What? Before our first anniversary? That sucks!" She shot back quickly.

"Oh, it's worse than that . . ." I paused with a snicker, wondering how she'd take the next piece of information.

"How!" she interjected sharply. We were cooking dinner together and with this additional information rolling in like a wave taller than her short stature, she dropped whatever utensils were in her hands, turned to me with her arms crossed, and braced for impact.

"Now it's a fifteen-month deployment, not twelve. I'm going to be away for our first two wedding anniversaries."

No reply. You could feel the wind being sucked out of the room. It was a combined realization that we wouldn't be together for our first two wedding anniversaries, and if fate would have it, possibly not any of them, ever!

Jenny fell into my arms as her eyes began to well up. She knew there was no reason to throw a massive fit about it. What good would that do? It wouldn't change anything. Sure, she was angry, and so was I. But nothing we could say or do would change the fact that we would spend this time apart.

The Army did one more solid: We deployed on June 30, 2007, to land in Kuwait on July 1st. Why is this significant, you ask? Good question. For whatever reason, combat pay was monthly and not prorated. So, if we arrived in Kuwait on July 20th, we still received a month of combat pay, even though we were only there for eleven days. I suppose Uncle Sam wanted to ensure we earned every penny of it. What pissed us off about this is that we knew Air Force pilots would use this to their advantage and fly into theatre for a few days to drop off resupply, collect their combat pay, and then fly back to home base without ever actually being in combat. Good for them, shitty for us.

Once in Kuwait, we had two weeks of briefings and vehicle rollover training to complete before heading to Iraq and getting started. I'll never forget one of the briefings, which familiarized us with a theatre-wide internet system that updated routes and enemy threats, allowing us to better plan convoys to avoid being attacked or blown up by known or suspected enemy threats and IEDs. The briefing was held in a typical conference room setup: tables were arranged in a T shape, with senior officers and NCOs seated at the top of the T, and lower ranks seated along the leg of the T in rank order. Even though I was an E-6 now, I still sat near the bottom as I didn't know who else would show up and didn't want to have to move. There were a couple of majors, a captain or two, some E-8s and E-7s, other E-6s from our company and me. I sat closest to the screen, but they had some nice leather chairs that leaned back, allowing me to see well.

In the middle of the brief, as I was somewhat leaning back in the chair, watching the screen and listening, my chair made a clicking noise and then leaned another inch to the right. *Maybe I should grab another chair.* No sooner did that thought cross my mind than the chair collapsed on the back of it. I found myself lying in the chair, on my back, on the floor. The room turned dead silent. Fortunately, they were high back chairs, so I didn't hit my head on anything. I kicked my legs, but the chair was still connected to the base and didn't move. I probably looked like an upside-down turtle at this point. I swung my legs to the side, and the chair broke free as I rolled to my side. Everyone was watching me. I'm sure my face turned red. I slid the chair out of the way as a buddy rolled another over. I sat down in the chair, strategically positioning it so that I wouldn't have to see anyone behind me, and, as it was still silent, I looked back over my shoulder. Everyone was quietly staring at me, and I said, "Well, carry on!" The room erupted in laughter, and the E-8 sitting across the table said, "Good recovery, Staff Sergeant!"

After the brief, I waited until everyone left the room before I left my chair. The rest of our company guys were waiting for me outside, and when I got out to them, we all had a raucous laugh about it. The story spread and it was used deployment long to break levity. "Is the deployment running long and sucks? At least you didn't fall over in a chair in a room with senior officers and NCOs!" "It's hot outside, you're out too long on a mission, or something you ordered didn't arrive in time? At least you didn't fall over in a chair in a room with senior officers and NCOs!"

After Kuwait, we headed to Tal Afar and Sinjar to continue efforts to shut down the border-smuggling operations. Other than that, it was PT and hangout until the next mission set would come down.

We had better living conditions than any prior deployment, as I had a CHU to myself, and there were enough housing units that we dedicated one to a team hang-out area with all the comfort items we brought. The FOB had a gigantic facility that had a well-equipped gym, theater room, full-size basketball court, pool tables, ping pong tables, video game and TV rooms, library, and best of all, an indoor computer and phone room that wasn't in a trailer or some other mobile platform. The dining facility was also the best I had seen in three deployments.

Amenities aside, missions started quickly, and our platoon was first in the chute. At first, it was primarily platoon-sized ops consisting of village engagements and surveillance, along with checking in on border guard checkpoints to ensure that business was on the up and up. Over four years into this, we were still doing the same kind of operations. It was a tad annoying, but at the same time, we were familiar with what we had to do and the area to do it in, so we hit the ground running and were ready for the challenges ahead. Being familiar with the area and ops was a relief, as we had a bunch of FNGs on their first deployment. We grizzled veterans could show them the ropes.

New people meant new officers, too, which was a bother since we knew the area, and the threat wasn't as significant as it was in larger cities like Mosul or Baghdad. That didn't stop the hype from passing around to the newbies. Intel officers would recommend not taking the same route twice, which is solid advice in locations with more predictable daily operations than what we were doing. If we rolled out one morning, it wasn't likely that we would be coming back the same day or even within twenty-four hours. Placing a roadside IED would only be a threat to the local populace, and even the insurgents didn't want to inflict harm on the locals as that would make the insurgency look bad. Regardless, our leaders took this intel to heart, and on one particularly long and strenuous day, we took a random open desert route to return to FOB Sykes.

It was night, we had single-optic NVGs, and the Iraqis in the desert areas were known for constructing berm barriers to signify land ownership or tribal land boundaries. So there we were, driving through what should have been a straight line to the FOB, which should have taken about an hour and a half, but due to these random berms, it took over four hours to return. We recommended that our PL take the roads back and save time, fuel, and hours, but it's rare for a new LT to take the advice of senior NCOs over other LTs. To drive around these berms we also had to pass through a garbage dump area south of Sinjar, near a small town called Baiji. It wasn't just a dump; sewage runoff went through the area, too, and we got our Humvee stuck in the sewage water muck that had accumulated a bit too high in one spot. We got lucky that none seeped into the vehicle and we could walk across the hood, roof, and trunk to hook up a line to be pulled back out. They spent so much time worrying about being tactical that we were put in a "fish in a barrel" scenario to avoid the unlikely possibility of a roadside IED. I've had plenty of circumstances like this in my military career where the voice of experience was ignored over the voice of the inexperienced, but college educated. A degree will never be as great as experience, ever.

Finally making it back to the FOB at 2 a.m., we did a quick unload and refit, hit the showers, and I was off to get some much-anticipated sleep. There was a knock at the door of my CHU, and it was the LT informing me that we would have an AAR in the bunker at 1 p.m.. It wasn't like a WWII bunker. It was a large, above-ground cement bunker that was our company AO. Since incoming fire wasn't much of a thing, the bunker gave us a protected place if incoming fire did become a thing. Anyway, I was still pissed off from the preceding night-into-morning events that I said, "Is that all?"

"Yes," he replied.

I slammed the door in his face, and he walked off. I learned later from our CO, whom I still greatly respect, that the LT came to him and wanted to submit an Article 15 for insubordination.

"If a senior NCO with numerous deployments and experience slammed a door in my face, I would look at myself and question what I did wrong to deserve such a response instead of looking to punish him for his actions," the CO told this LT. I greatly respected the leadership of our CO and how he looked out for us senior NCOs.

These missions were the norm for the first third of the deployment. When we weren't in the field, we all established routines based on the TL's discretion. I didn't want to exasperate my team, so we did team runs twice a week on the marked 5K dirt route near our living area. The dirt track was also used for monthly fun runs or competitions that the FOB would host to keep morale up, such as Halloween or Thanksgiving events. There would also be prizes and T-shirts for those who participated, but we never did. At least not my guys. Other than that, I let them have free rein of their days if we weren't going to the range, servicing our vehicles, or conducting any other requirements necessary for the day. Operations were conducted and initiated at night, so we were on a reverse schedule for the duration of the deployment. I liked this schedule since midnight chow had breakfast, which meant I could end the day with some egg bowl and toast, my favorite.

My typical day was getting up around 10:30 a.m. to 11:00 a.m., getting dressed, and brushing my teeth. I'd see who else was up, catch the bus, or load a truck, and go to lunch. We had daily Battle Update Briefs (BUBs) at 1 p.m., after which we would pass out all pertinent information to the rest of the team. We would conduct whatever training we had planned for the daytime, go to dinner at 5 p.m., workout at 7 p.m., make phone calls or check emails at the multi-use facility (MUF), and then conduct any nighttime training we wanted to. If there weren't any nighttime training, we would all hang out in some form or fashion, watching a movie or playing a video game, have midnight chow, and then off to bed. Deployments, like prison (I've heard), are all about establishing a routine to maintain your sanity.

Leave was coming around, and since all my guys wanted to be home for an anniversary, birthday, or some other special occasion, I chose to go home for Christmas and the New Year. No one had an issue; everyone had a reason for their leave dates, so I took the holiday season. December came quickly, too quickly. We were six months down but by the time I would return from leave I would still have nearly nine months left of the deployment. I thought that maybe I should have waited a few more months, something in the middle of the year like Jenny's birthday in May, but I was already in Kuwait and on my way home. It was too late to change plans. However, being home for the holidays was a good tradeoff for missing our first two anniversaries.

Being finally at home with my wife for Christmas was fantastic. We had dinner with her family, who mostly lived in the area, and it was always a big to-do. There was plenty of food and family around, which I had been accustomed to growing up and was grateful for. Some friends of ours who had gotten married while a bunch of us were deployed decided to host another wedding reception for friends and family who couldn't attend their original wedding. We met up with all of them in Winston-Salem and had a great time catching up before we headed off to a surprise my wife had planned for me. We went to the Biltmore in Asheville for New Year's, and it was a sight to behold. We toured the mansion—which wasn't a castle but might as well have been—and the gardens. Then, for New Year's Eve, we had a planned five-course meal with wine pairings.

I decided to partake in the pairings—after all, I had been sober for two years now and felt like I could handle having a drink now and then. There was just enough wine that I had a nice relaxing buzz going, but not enough to get drunk and regret the decision. We also attended a wine tasting and purchased a couple of bottles of wine to take home with us.

Before heading back to Iraq, we drank most of the rest of the bottles we bought, but again, I didn't get drunk, which boosted my confidence that I could maintain control while having a few glasses of wine with my wife. If I could keep my alcohol consumption under control now, I could surely do it in the future.

CHAPTER ELEVEN

B reak time was over, and it was time to return to the grind. Our unit had moved from FOB Sykes in Tal Afar to FOB Speicher in Tikrit. I hadn't emailed anyone to let them know that I was returning. I was pleasantly surprised when we arrived at the company AO to find we had new living quarters, more Gators and pickup trucks since everything on the FOB was further spread out than what we were used to.

We also had brand-new Mine Resistant Ambush Protected (MRAP) vehicles for our patrols. They were like up-armored rolling buses. We had plenty of room for our six guys and gear. They also came equipped with internal communications headsets so we could talk. We typically had one headset with an earbud connected to the mouthpiece to listen to music while rolling. We looked forward to having them in the summer as they had a robust AC system. Lastly, a V-shaped hull made them less susceptible to being blown to pieces by a direct IED hit underneath the MRAP.

New gear aside, there were suspected terrorist activities in the area, and the command team was on top of getting our guys in on the action. At long last, we would conduct team-level missions on a shorter duration of thirty-six to forty-eight hours. These missions were to be "eyes on" surveillance before a larger element performed a raid on a village where intel stated insurgents had been meeting to plan for their terrorist or guerrilla-type actions against allied forces in the area.

Our first mission was to surveil an old rundown salt mining factory beside an enormous salt flat in the desert. I attended a briefing with the aircrew as we were inserting via Black Hawk and thought some of the questions were hilariously stupid. They asked me: "Which heading do you want us to fly in on? Which heading do you want us to land? Where do you want our gunners oriented on landing? Do you want us to loiter?"

While I didn't fully understand the reasoning behind these questions, I answered them as best as I could as they came up. I finally halted the briefing—which, by the looks on their faces, doesn't happen often from an E-6—and had them pull up the slide with the LZ on it. I said, "I want to land there at that 8-digit grid. I don't care how we get there, and I don't care which way we are facing. I want to go *there*. Once we land and the crew chief gives me a thumbs up, we will hop off with a three-step drop so you can fly away. We won't be in the gunner's line of sight so that he will have a clear field of fire. Once you take off, I would like you to loiter approximately ten clicks away to the north for ten minutes in case we make contact. After that, we are on our own. Any questions?"

There was a short pause and then laughter, followed by, "Whatever you want, Sergeant. We got you!"

Then one of the crew chiefs asked, "Can you turn around and close the door after you get off the bird?"

I give him a sarcastic eye roll look and said, "Sure." A bit more laughter, and the mission was a go.

For this thirty-six-hour mission we didn't pack much of anything. Over the years, we upgraded tech somewhat to lighter-weight Toughbooks, along with batteries that last longer while having more connected to them and fantastic camera lenses with lightweight adjustable tripods. We took just enough cold weather gear, food, and water to complete the mission, plus one day. That was the lightest combat ruck I had ever had to endure.

The Black Hawk insertion went off without a hiccup and we moved to our HS. We didn't split into two sites since it was a wide-open desert with no actual cover and only tumbleweed brush for concealment. We found the lowest ground with a lot of scrub brush and decided to make that our HS. It was about a click from the factory to our south and the same distance from the salt flat to our east.

The sun rose, and we were all set with our equipment. We had a decent spot to hide out, and with the whole team, we were ready for Easy Street: Get some rest, take some pictures, and send them back; pull security; twenty-four hours remaining. Too easy. We didn't see any vehicle activity, only one sheep herder. It's always a random sheep herder that would be the cause of compromise, but not this time. He never came close enough.

By midday, we sent up some photos of the factory that the guys in the rear turned into a battleship diagram. Every door and window was labeled for ease of future operations in reference to entrance points or points of interest for potential enemy fire. *What about this sheep herder? Is he the mastermind? Is he moving in under the cover of being a sheep herder and setting up the following insurgent missions? Maybe he's the cash guy handing out payment for missions completed. Nah, I'm getting tired; he's just a sheep herder.* Still, we were ready, willing, and able to light him up if he was more than that. He wasn't.

We finished up, and the intel was off. No one was meeting up that day, and with follow-on non-human surveillance, no one met up there the following days. Still, it was nice to get out with the team and ensure we were ready to meet the needs of human intelligence (HUMINT), which was requested by those higher-ups who still valued it.

Our next and last team mission was a huge build-up to a massive disappointment. We had intel on known insurgent activities in a village and confirmed meeting locations for planning and staging. This mission was the one—the big show for real. We were going to see something worthwhile for sure. We were going to be the spearhead that crippled the insurgency. It was thirty-six hours of eyes-on before an infantry company came in full force with mortars and extra assets to destroy any insurgent fighters. They would arrive at 9 p.m., landing with three Chinook helicopters. We would brief their commander on any last-minute updates, and they would storm the castle while we moved off, got picked up by a Black Hawk, and returned to base, heroes of the day. Or that's how it sounded before it started. In the end it was a gigantic nothing burger. There was only a small family living in the village, consisting of abou four mud hut houses. No insurgent activity, no meetings, planning, or staging equipment. Just major disappointment.

Our time at Speicher was short-lived, lasting about three months, and we were packing up to head back to FOB Sykes in Tal Afar. I was astonished when we drove through Mosul, and what was once a bustling city now looked like it was on lockdown, even during the middle of the day. There were giant holes in walls that had to have come from tanks, broken-down vehicles on the side of the road, and buildings with missing walls. All in the same area we patrolled nearly five years earlier, with swarms of people everywhere. The guys who had driven through while I was on leave told me about it but seeing it in person was dramatically different. It was rather heart-wrenching to see the effects of war but so much more so to know what it was like beforehand, not even from pictures but from having been there. I recently looked on Google Maps to see some locations, and they have since recovered. There was a shopping mall and restaurants I know weren't there before, but still, at what cost and to what end? War is hell, but the recovery effort brings new life. I'm not advocating for war; I'm just making an observation.

Tal Afar had the same setup and location: CHUs, village engagements, and traveling desert life. I had hit my peak as a TL and was ready to pass the hat to my more-than-capable ATL. He didn't want to take a team, but it was time. During our time at Speicher, two platoon leaderships were fired. One decided to drive back to Speicher on a route labeled black while they were out on a mission. (Black routes signified a known IED threat; do not drive.) They ignored the orders and went out anyway, resulting in one of their vehicles being destroyed. While no one was KIA, the gunner suffered a traumatic brain injury (TBI). Another platoon leadership got relieved, which left one opening for a PSG and PL that I did *not* want any part of. I asked the CO if I could "ride the pine" for the rest of the deployment in S3, but he had a counteroffer.

They offered me the PSG position. I was tired of rolling around the desert and even more tired of dumbass LTs. No thanks. They sweetened the deal by stating that Derreck would be the platoon leader. It would be an NCO-led platoon. I went to Derreck's CHU to talk it over with him.

"Did they talk with you?" I asked, figuring he might know what I meant.

"Yeah," he said.

"What did you say?" I asked.

"I said OK."

"Then I guess I'll tell them OK too."

And I took over first PLT, as easy as that.

One of the CPTs accompanied us on our first few missions. He was our PL before he was promoted, so we all had worked together before. He mainly observed our conduct, which would be considered our training wheels before we were allowed out without adult supervision. It all became a routine: roll out, talk to village leaders, determine their stance on the insurgency, report it up, and map out the entire western side of Iraq near the Syrian border.

We had five to six more months of this shit. What made matters worse was my back, which was starting to show signs of wear and tear after these numerous deployments, Airborne ops, wearing body armor, riding slumped in military vehicles, and carrying heavyweight rucksacks. I would randomly throw my back out just picking up my boots or something light and be in pain for the next few days. One day, I was working out with Derreck, and my back popped so loudly he heard it ten feet away from me. I went to the doc, and they didn't hesitate to give me pain relievers. I spent the rest of the deployment popping Flexeril like Smarties candy, and that still didn't help the pain all that much. I had that back pain for the next year without anything giving me relief: no pills, no physical therapy. I was not excited about the missions ahead, knowing I had to hide the pain as the new PSG.

My mission-day routine consisted of waking up, having a giant cup of coffee while trying to stretch away the pain, waiting for the Flexeril to kick in, prepping all my gear, loading up with the platoon, and rolling out. I would have to sit with a wincing, pained look until my back would settle into the new position of being in a seat. Once I got used to sitting, it was OK. Each bump would send a stabbing pain from the lumbar spine area, up my back and down my legs, then would return to my lumbar region and just bounce around like that until it subsided. When we would stop, I would hobble out of the back of the MRAP, walk it off a bit, and then get used to standing and walking. I was hurting *way* more than I ever let on, but I made it through OK.

Now and then, one of the TLs who would notice me moving a little slower or stopping due to the pain would check on me privately. They knew but never made a spectacle out of it, and I appreciated that.

We had a moment right after we took over the platoon where some teams were messing up. I don't remember what happened, but Bliz pulled Derreck and me aside and gave us a friendly but firm pep talk. He politely crushed our nuts since we were friends, but that didn't stop him from making sure that soldiers were conducting themselves as soldiers, not shitbags. I told Derreck I would handle it, and after one of our afternoon meetings, I started to tear into the TLs. At the time (and even today), I didn't swear very often as it would have a more significant effect when I did. So, after telling them to stop fucking up and make sure their teams were conducting themselves appropriately, I sent them on their way.

A few days later, Sammy, who was 235-pounds, six-foot-three-inch tall, former Division 1 football player for the University of Houston, came up to me and said, "Damn, Sergeant Kruger, when you were tearing into us and started swearing, you scared the shit out of me!" Keep in mind, I'm five-eleven and about 175 pounds so I was surprised to hear that I scared *him!* I told him my trick about swearing only when necessary, and he said he would try that one.

Operational tempo (OPTEMPO) picked up considerably, and we were the go-to platoon. We were out on a mission for three to five days at a time and only back for thirty-six to forty-eight hours and right back at it.

We came across an abandoned village near the Syrian border, which was about six to eight mud huts, some short-walled hut with no roof, probably a large pen for sheep when they weren't grazing, and that was about it. The sheep pen was sand-filled, which was odd even for the desert as somebody had piled the sand relatively high. I had one of our guys dig around, and soon enough, we found some large bottles and tubes. It was an acetylene torch used for constructing IEDs. I ordered the men to spread out and check all the buildings, and as I walked into the next one, the floor gave slightly under my weight. Painfully leaning down with all my gear, I saw the corner of a HESCO barrier, a thin, collapsible wire barrier with burlap linings that can be set up quickly. They get filled with dirt or sand, and voilà, you have a strong and sturdy barrier.

I had two guys grab the corner and pull it back. Insurgents had used our barriers to make a fake floor to cover dozens of bags of homemade explosive (HME) materials. As I stepped out of the hut, I hollered at the rest of the platoon to take a look, but a few of the guys in another hut said they had the same thing. After checking every hut there, we had over twenty, 50-pound bags of HME, the torch, a cement mixer, and a bag of cinnamon. After collecting all of this and contacting the CP, they were able to get a Black Hawk out there with an EOD team to set charges and blow it all up. We piled everything up in the corners of two adjacent huts, and EOD strapped 100 pounds of C-4 and detonation cord (det-cord for short) to it and lit the fuse.

All men and equipment moved back to a safe distance. Anyone with a camera got it ready and held an NVG in front of it as it was dark out already. The fuse was lit, and they loaded up to meet with us. "Five minutes. Mark," came over the radio. We sat back and waited for the show. Once all together, the EOD TL counted down the minutes. Three more. Two left. One more minute. Time. No explosion. All eyes were eagerly watching. About thirty seconds more, then *BOOM!* Watching it in real-time was surreal. We saw the explosion, but light travels faster than sound. Without NVGs on, we could barely make out a shockwave, but within a few seconds, the sound hit us, rocking us and the vehicles, and it sounded like someone was standing right behind you shooting a double-barreled shotgun off, both barrels at the same time. It was a deeply satisfying feeling to know that our efforts and diligence had saved the lives and limbs of Allied forces. I called it a career bust. This also helped shake the lethargy of a long deployment and motivated us to look even harder on every mission, regardless of how routine everything began to feel.

We didn't find anything else like that on subsequent missions. With such a high OPTEMPO, we informed our company leaders that the other platoons needed to pull their weight, so we could spread the missions around and get a few days of downtime in between. We had a bunch of random happenings in the middle of the desert, which were comical, to say the least. In the middle of nowhere, we came across an old dirt bike. One of the guys asked if we could load it up and take it back to ride on the FOB.

"If you can get it started, then yes."

It wouldn't start.

So, what do a bunch of boys with big guns want to do next?

"Can we shoot it?"

"Hell yeah!" I replied. With a 50-cal full of API (armor-piercing incendiary) ammo mixed in, they took a three-to-five-second burst at it, and when the API round hit the bike, it burst into flames. Wanting to see more, they took a few more shots at it, and the fire followed the bullets as they went through the bike. It was something straight out of a movie's special effects magic.

Another abandoned village had a mud hut about the size of a large master bedroom full of tumbleweed.

"Can we light it on fire?" someone asked.

"Seems like it would be a waste *not* to!" I said.

We set that sucker on fire and rolled out. We were a mile away when we saw the smoke plume. It wasn't near any other hut, and the fire sure wasn't going to spread through the desert. Best we could tell, the place had been abandoned a long time, and no one was coming back. *Burn it all for all I care.*

A random broken-down water or fuel truck in the desert? Blow out the tires and make it completely unusable! Smiles on faces that would have otherwise been home already. At this point, we were past the twelve-month mark, and every day was a twist in the knife of our hearts. These little moments of levity were among the few that brought laughter and smiles to our faces—blowing off some steam.

This would be the last of our missions in Iraq. I never shot a single round at an enemy in a time of combat. I made peace knowing I wouldn't be this killing machine or warrior on the battlefield. I guess God had other plans for me. This deployment was brutal, to be honest, but something about the company we had seemed to make it not so bad: camaraderie. It changes things. If we weren't out on a mission, we were all hanging out together, discussing families, plans, and what we would waste money on when we got home—smoking cigars, listening to music, watching movies and TV shows, and playing video games, drinking coffee late at night and early in the morning, talking, and more talking. I know these guys better than anyone else in my life.

There's little more to tell about these fifteen months of drudgery and misery. It was finally time to go home. Traveling to Kuwait and home should've been easy, but, alas, it was not. Murphy showed up and hit us with another three-day sandstorm, similar to the one in '03, so we were stuck in Baghdad for three days, which pushed back every flight in and out of theater. That meant we stayed in Baghdad for five days before we finally caught a ride to Kuwait. This also pushed our flight out of Kuwait back, and we were informed it would probably be another week until we got a flight to the States.

Fortune smiled and we only stayed three days in Kuwait. "Grab the troops; we have a plane; we're going home!" Bliz yelled with a huge grin, the CO right next to him. We all hopped on the buses to the airport, where they pulled up next to a 747. Allegedly, the plane was designated for another unit, and we weren't supposed to be on it, but Bliz looked at the dude who was in charge of it all and said, "Our guys are getting on that plane, and we are going home," with a solid 1000-meter stare into his soul. He handed him a copy of the manifest, turned to us, and said, "Get on the fucking plane!" I had four seats to myself the entire flight, and even though I could lie down across them, my back pain overrode my ability to sleep, no matter how much Flexeril I took. From Kuwait we flew to Germany then to Newfoundland (for whatever reason), to Chicago, and then to Bragg.

On September 15, 2008, we finally arrived home, having been away about two weeks shy of fifteen months; it had been nine months since I had seen my wife. I had plenty of leave, so I took thirty days again to grow a beard and not worry about wearing a uniform. Plus, my wife and I were going to Kauai, Hawaii, where we had a blast. While I was away, Jenny put everything together as we planned it. We stayed on the north side by Hanalei Bay in a beautiful resort with a vast pool of three connecting parts. We decided to do four activities during the week, giving us a few days dedicated to exploring the island and taking pictures. We took an inner tube tour that floated along an old irrigation canal used by the sugar cane farms, (I don't think this tour is available anymore). We did an hour-long, doors-off helicopter tour on an MD-500 helicopter. It flew so close to some of the waterfalls that we were getting misted with the doors off, which only whet my appetite for going avia-tion. Of course, we attended a luau that featured a fantastic show of Polynesian dances at the end. Our favorite activity, which we have now done twice, was a four-hour ATV tour.

The ATV tour was a blast as the tour guides would halt the group before any major puddles or water crossings, inform us where to cross them to maximize the muddy carnage, and let us go at it to have fun. They gave out bandanas so you could keep the mud chunks out of your mouth and wipe the goggles off, but it ended up like being water-boarded with mud. You were going to taste it one way or another. Mid-tour, we stopped at a small waterfall with a pool of water beneath it, and since it was rain-fed daily from the mountaintop, the water was cold, especial-ly in late October. We could jump off rocks into the pool of water and swim a bit before the provided lunch, and then it was back to more sugar cane farm roads, mud, and famous movie sites. I'll never forget that trip to Kauai. It was a great way to celebrate after missing our first two anniversaries.

While I was on leave, the military battalion had a change of command from top to bottom, and the new ser-geant major was ready to shake things up. After I returned from Hawaii and tried to get back into the groove of being at work, the SGM came in during a company meeting to ask us how long we'd been in the unit. For anyone who was an old-timer like me, the SGM would tell them, "You're outta here" or "You're going away."

When I told him I'd been in for eight years, he told me I'd become a Ranger instructor. I wouldn't have had much of an issue, but my back was in shambles. Any thoughts about trying out for SF or anything in the Special Ops community were out as I didn't want to be mid-tryout, throw my back out, and follow that with the walk of shame as I would have to Drop on Request (DOR). After he left the room, I said,

"Looks like I'll be putting a rush order on my Warrant Officer Packet."

After completing all the paperwork for the packet, I sent it through the appropriate channels, and then I had to wait. I waited four months to learn that I had been selected and received orders to attend Warrant Officer Candidate School (WOCS) in mid-September 2009. I was going to fly helicopters and was ecstatic to hang up my ruck. It was going to be a better life, or so I thought.

Jenny was four months pregnant, and the pregnancy was sapping all her energy. Neither one of us wanted to deal with a move, but here we were, two vehicles packed to the hilt and our household goods being shipped. We arrived at Fort Rucker on August 10, 2009. A new way of life, and I would learn rather quickly, a new Army. At least not an Army life I was aware existed. I had a few weeks of leave before starting WOCS, so we settled into our new location for the next two years. We had frequented Wrightsville Beach and Myrtle Beach while in North Carolina, so we decided to take a trip to Panama City Beach and Destin Beach. For those who haven't been, the Emerald Coast is amazingly gorgeous and rivals some fantastic beaches worldwide.

I was feeling pretty good one week before I started WOCS, so naturally, I threw my back out. The first thing in WOCS was a PT test. I was nervous about it because I had messed up on PT tests before, even when I didn't have such bad back issues, and I was already starting to wonder what would happen if I failed this by not being able to perform well enough. I let fear of failure dictate my feelings, actions, and reactions. This time was no different. Wanting to get out of my head, I grabbed heat and ice and went to work, rotating between them. I also took a lot of Motrin and went for some walks with Jenny. I discovered a smartphone yoga app with two different routines you could tailor to your desired length. I chose thirty minutes and did that once daily in the morning. I was surprised that my back felt a bit better after five days, and I was able to pass the WOCS PT test and WOCS without issue.

December 31, 2009, Jenny and I are watching the start of Dick Clark's New Year's Rockin' Eve when she announced that her water had broken. We had already packed a bag, as the baby was due within a week, so we loaded the car and headed to Flowers Hospital in Dothan, Alabama. Unsurprisingly, there was next to no traffic at 9:30 pm on New Year's Eve, and we arrived at the hospital without any pregnancy-related issues. No traffic during the drive was great since I'd had a couple of drinks, but I felt fine since we'd been snacking on all the finger foods we had bought for the evening. Still, it didn't sit easily with me since I'd had a DUI in my lifetime, and now I was a warrant officer and "should know better."

We got to the hospital, they hooked Jenny up to every machine possible, gave her the epidural, and eventually, we were in for the waiting. We celebrated New Year's with the nursing staff there and continued waiting. Jenny was ready, but the baby was staying put. Around 9:30 a.m., they decided to induce labor and gave Jenny some Pitocin. That did the trick. At 10:10 a.m. on January 1, 2010, Charlize Esmeralda Kruger was born into the world. Jenny and I chose the name Charlize because we both loved it, and Esmeralda was her grandmother's name. I also love it because Charlize and I have the same initials: CEK. She is the second most beautiful thing I have ever seen, and we welcomed her with open and loving arms.

Soon after her birth, I had to get back to work and start SERE School, a requirement for aviators as the potential to be shot down and captured in combat was a greater possibility than the average grunt just randomly getting caught from their platoon. Unfortunately, we were required to sign a non-disclosure agreement, which prevents us from discussing what we learned there. It was some fantastic training, and how they taught it should be the model for all course instruction. I learned a great deal about my mental capacities to cope and endure if I were ever to become a POW or detainee, which I am grateful to have learned. Oh, and no, they didn't break any bones, but I did get slapped in the face a handful of times. Pun totally intended.

There was one night that was quite comical. We were nearing the end of the SERE School, and there were a few days in the field where we practiced our survival and evasion skills before being picked up by "friendlies" in a white van. I can say, never get into any white van! I don't care if you know the driver and all the occupants. Don't. Get. In. The. Van! We had a night to wait before the white van came to pick us up, but we had been sent to the same location as other teams, so we had about a dozen folks with us now instead of just our team. All we had to keep warm was a poncho that doubled as a carry-all during our time in the field. As some of us were trying to find a soft spot on the ground with some vegetation, one of the guys said, "You were infantry. You're a Ranger?"

I said yes. He asked, "You want to spoon to stay warm?" I laughed and said, "OK." As I finished saying that, someone else said, "I want in on that!" Within a few minutes, we had six guys, all nut to butt but warm with five ponchos on us and falling asleep within minutes.

In the middle of the night, I woke up, and since I was still experiencing back issues, I announced that I needed to roll onto my other side. In unison, we all rolled over and returned to sleep in no time. We all slept better than if we had just shivered the night away individually.

June 10, 2010, at nearly the ten-year mark in my Army career, I finally got the chance to get behind the controls of a TH-67 training helicopter. I had been looking forward to this for the last year and a half since I started my flight packet—actually, for the previous seven years when I saw those Kiowas buzzing around Iraq. It was happening, and I was excited, but my excitement would be short-lived. I sucked at flying, and I wasn't progressively getting any better like my flight school companions were.

A couple of weeks in, I couldn't hover much better than I did the first few days at the controls. I could fly a standard rate turn in a climbing or descending corner of a traffic pattern, maintain altitude in straight and level flight, and hold a constant airspeed within the standard. However, my hovering, landing, and takeoff all left much to be desired. Emergency procedures? I could barely focus on remembering my name, let alone what to do when a simulated emergency was announced. I wasn't getting it, and it felt much worse than jacking up an exit in HALO school. I was ready to quit. Everyone, and I mean everyone in the class, was getting it all quicker than I was. *Am I even cut out for flying?*

Sometime around mid-July, after I had told Jenny more than once that I was ready to quit, she finally said, "Then quit! Or don't! But make up your mind and do one!"

I committed to the road we were on, and while my hovering with that fickle bird never became great, I did get better and passed the first phase of flight school. Not with flying colors (more puns), but I passed above average, and since I owned the above-average ground, I was perfectly fine with that. I was moving on to instruments.

Instrument flight can be fun, but not so much if you like looking around and seeing things. After all, instrument use is supposed to be for when you can't see because of clouds or when it's nighttime. Most of the instrument training takes place in the simulator, so that was a nice change from being hot on the flight line in Alabama in August. I did well in instruments but was nervous about returning to the flight line after six weeks of not being in an actual helicopter. The instructors did all the flying while we adjusted the instruments and instructed them on headings, altitudes, and airspeeds. Listening to the radio was a significant part of Instrument Flight Rules (IFR), as we would take flight directions from flight-controlling agencies while flying. It became somewhat routine, and since we were inundated with it for two months, I was sure I was ready when test day came.

The IFR test came, and my stick buddy flew first. They were gone for a little over an hour. I used the time to review any information I thought was relevant to my flight. They landed, we swapped, and I set the instruments for departure from Cairns. I performed the takeoff, and after air traffic control (ATC) provided us with our heading, I transferred the controls to the instructor, briefed him on the plan, and set the instruments for the task ahead. I informed him of the heading, altitude to which we'd level off, airspeed, and the intercept course for our next heading, just as I'd been taught to do. Students were to set the parameters and have the instructors fly, and we were good to go. I conducted our level-off checks and started a fuel check. I was ahead of the aircraft, and life was good.

A few minutes later, the instructor said, "When did you want me to track that outbound course?" I looked down and saw he had wholly blown through the course. *Did I ask him to intercept?* I was sure I did since he leveled off at the altitude I instructed him to. *Did he blow through it to see if I was paying attention or if I would catch it in time?* Yes, it was my fault, but we had been taught that on test day, the instructors would do everything we told them to do and would be competent co-pilots. He was proving that wouldn't be the case. I panicked, took the controls, and got us back on course as quickly as possible without losing altitude or airspeed. I returned the controls to him and set the instruments to enter a holding pattern. I tried to shake it off as best I could, but I questioned whether I could trust this guy or if that would be enough to fail, and I saw my aircraft pick slowly flying out the window. Punny.

We made it to the intersection holding, and I adjusted the outbound flight length for one-minute turns. I was feeling better about it, too, as I didn't use the GPS even though it was an option—I decided not to use the GPS since he would have had to set that up, and I wasn't sure if I could trust him. A couple of turns in holding, and it was time to depart. ATC gave me the heading, and we went to Enterprise Municipal Airport for the VOR approach. A few tasks were completed, and now I had a non-precision and a precision approach. Easy. I checked the instruments set for the very high frequency omnidirectional range station (VOR) approach and briefed the instructor on the procedures: "I have the controls."

Somewhere in my frustrations and anxieties, I had missed that it was a step-down approach. A step-down means that at some point you descend to a particular altitude, wait for a distance or time, and then step down to another altitude or the final altitude from which you would land. I was too high and realized I needed to descend. I dropped the collective and almost initiated an autorotation to reach the final altitude on the approach chart. Luckily, I was within the allowable deviation of +/- 100 feet when we crossed the threshold for the approach. While it wasn't pretty, and my confidence took a hit, I did pass the non-precision approach.

I had no doubts about my ability to nail the ILS approach into Cairns Army Airfield, which was the final maneuver. I set the instruments for the approach, and ATC was vectoring us to the final landing course. I put everything that had happened up to this point out of my mind and focused on the approach. I nailed that ILS! I mean, I nailed it! Every pointer that showed my altitude, airspeed, and rate of descent was locked in. I flew that thing down to the last foot and called visual upon seeing the runway. After that, the instructor took the controls, and we landed, shut down the bird, and went in for our grades. Nervousness, tension, anxiety—thy name was Chris. We sat down at the table and went over the entire flight. After discussing the good, the bad, and the ugly of my flight, the examiner gave me a passing grade as he didn't think more flight time would be necessary. My grade reflected accordingly.

It was time for Basic Warfighting Skills (BWS)—and time to have fun in these pieces of crap aircraft For BWS, we got an upgrade. Instead of the TH-67, we would be flying the OH-58C, which was easier to control due to longer blade lengths and tail booms. I was surprised to see how much a minimal change to just a couple of components made to the aircraft's flight characteristics. It was fun, almost a joy to fly! I was chomping at the bit even more for my potential future as a pilot of one of these birds. BWS was more geared toward tactical flight than anything else. Before we started, we had to put together the infamous and dreaded map book. We were all given thirty maps of the surrounding area, which had to have the margins and marginal data removed. Next, we had to paste these together in a folded manner that allowed us to flip through the pages like a book, enabling us to fly our routes without losing track of where we were and where we were going. Finally, it had to have all the obstacles and hazards up to date from the online library. Updating the obstacles required us to hand-draw them in appropriate colors on every map page. Our class was fortunate to have a four-day weekend before starting BWS, and I spent most of that weekend working on this book. I poured so much time into it that I still have it on the shelf in our office. Since I had plenty of military map experience, conducting aerial nav wasn't more challenging than walking. The only difference was moving at ninety knots. If you had good visual references, it wasn't difficult to stay on course, and if the person flying was decent, it was easy to keep on time. We had to plan and fly a route every day. After completing that satisfactorily, we were allowed to "play around" with the flying. I'm grateful for the instructor I had during BWS, as we played a good bit. I only got off course once, and he said, "Oh! Even the great navigator gets lost now and then, too, huh?" We laughed that one off and still had fun flying low-level down in the trees and above the riverbeds. The hover work in fields and around obstacles, amongst other maneuvers well within the aircraft's capabilities was worth teaching. Ultimately, he put me up for a 98 percent, and I got a 98 percent.

There was only one more graded event between us and the airframe selection—a final graded PT Test. I was glad the PT test was in mid-November, as it was 55 degrees. I scored in the mid-270s on the PT test, so I had hoped that would be enough to get me into the first pick, but then I heard that one of our guys scored 350 or something around that on the extended scale. It was over if he had a better instrument exam score than I. A couple of days later, it was time to select an airframe.

Up to this point, I had busted my ass studying to make sure I would get the first pick of aircraft After completing Initial Entry Rotary Wing (IERW), which included the first two months of traffic pattern work, two months of instruments, and one month of Basic Warfighting Skills, they held an airframe selection day. The higher your scores were, the lower your number would be to pick an airframe. We had all heard that if the first pick didn't get the airframe they wanted due to availability, the battalion CO of flight school would ensure they got what they wanted for both the officer and warrant officer. Everything was taken into account: classroom test scores, flight exam scores, and PT tests. I was in a class of intelligent people, most of whom had some prior aviation flight experience in the civilian realm. Some were previous crew chiefs who had already been exposed to aviation, so it wasn't completely foreign to them as it was to me.

We all headed into a classroom, knowing the deal from the students ahead of us: we go in, and they announce the pecking order and available airframe numbers. Officers select first, WOs choose second, and the National Guard gets whatever their home station uses. I discovered I was the third pick, but I still felt good about my chances. I knew the first pick wanted Kiowas, and the second pick wanted Chinooks, so if there were two Kiowa slots, I was in!

They had the aircraft and available slots written on a whiteboard, and before they turned it around, they asked, "Who wanted attack?" A few of us raised our hands, and then they turned the board around. I'm sure my jaw, as well as others, hit the floor when we saw that there were no Kiowa or Apache slots at all! Three UH-60M, five UH-60A/L, and five CH-47F model slots, that was it! They asked the first pick what he wanted. He sarcastically replied, "Kiowas!" They told him to pick one on the board and he chose Mike Model Hawks. They erased a mark and moved to the second pick.

At this point, my stick buddy leaned over and said, "Take the Mike Model. It has a glass cockpit, all digital." The second pick chose Chinooks.

My turn. "Mike Model, I guess," I said as I lifted my hands in defeat. I shit you not, I wanted to walk out right there. I wanted to say, "Send me back to the infantry!" But I accepted my fate. All that work to get an airframe I didn't want. I would be stuck with Black Hawks for the rest of my career. I had worked hard my entire military career, and everything I wanted to do had just come to me. I never made it to Ranger Battalion to become like Cameron Poe, but in turn, I got LRS, which led to some of the best and most enjoyable training I had ever participated in.

Blinded by disappointment from lack of personal choice fueled my frustration. I wanted something with guns on it and a primary mission of destroying the enemy, not an aerial taxi. I didn't care who was in the back. I never did. I always felt like a taxi, and what made matters worse later was the higher ranking the individual in the back was, the more they *looked* at us like taxi drivers. The final twist in the shiv that was in my side was knowing the class before us had a few Kiowa slots and a bunch of Apache slots. We heard later that the class after had the same. What made matters worse for me was that after airframe selection, the BN CO came in, grabbed the guys who were first pick, took them to the back of the class and asked, "Did you get the airframe you want?"

Both, "No, sir."

"What did you want?"

Both, almost in unison, "Kiowas, sir."

"You got it, go sit back down."

Another guy starts to ask as the BN CO is walking out, "Sir, can I get a . . ."

The BN CO cuts him off without even looking back, "Only the first pick."

It's then I knew I was going to be stuck with the Black Hawks for the rest of my career. I took it as a sign that I was supposed to fly hawks, but that didn't dissipate my disappointment.

Out on mission in Iraq, 2008.
Rest in peace Bliz and Sammy (the two on the left)

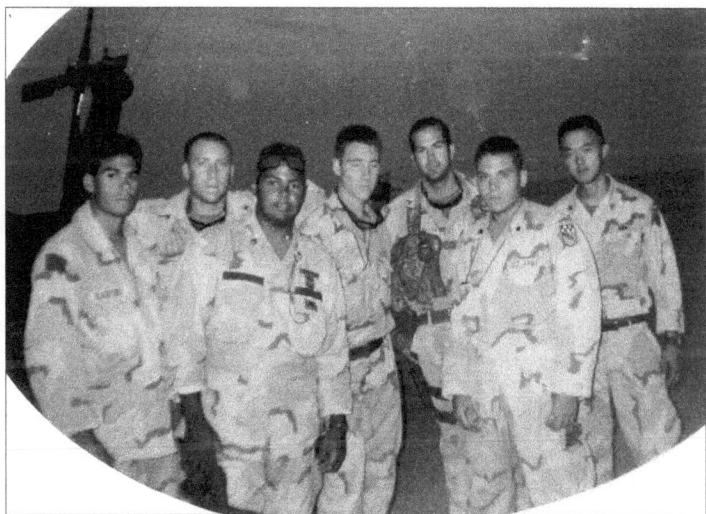

After desert training near Q-west in Iraq, 2004

3rd Platoon in Mosul, Iraq, 2005

3rd Platoon in Tal Afar, Iraq, 2005

*Chris and Genevieve, two months into our relationship.
Deploying that day.*

HALO operations, Raeford DZ, 2004

Reflexive fire training. Iraq, 2004

Days in purgatory counted. Mosul, Iraq, 2005

Failed assassination attempt. Purple Heart Alley, Iraq, 2005

Fast Rope training circa 2006. Fort Bragg, NC

Glamour shots at FOB Shank, 2012

C Co, 5/101 at FOB Shank, 2013

11,500 feet over a mountain in eastern Afghanistan

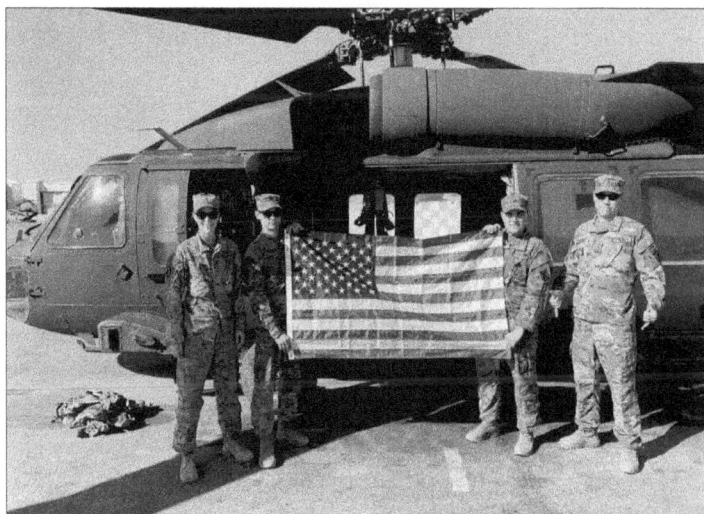

Christmas Day 2019 in Kandahar, Afghanistan

Flying around eastern Afghanistan, 2015

Flying around the Lanai shipwreck, 2017

Lotto Tower, South Korea, on a smog-free day, 2018

One of many "meet and greets."
Some desert village in western Iraq

Snow angels in hell. Iraq, 2005

It snows in hell. Iraq, 2005

F Co 51st LRSC Company coin painted flawlessly
on a T-barrier, Iraq

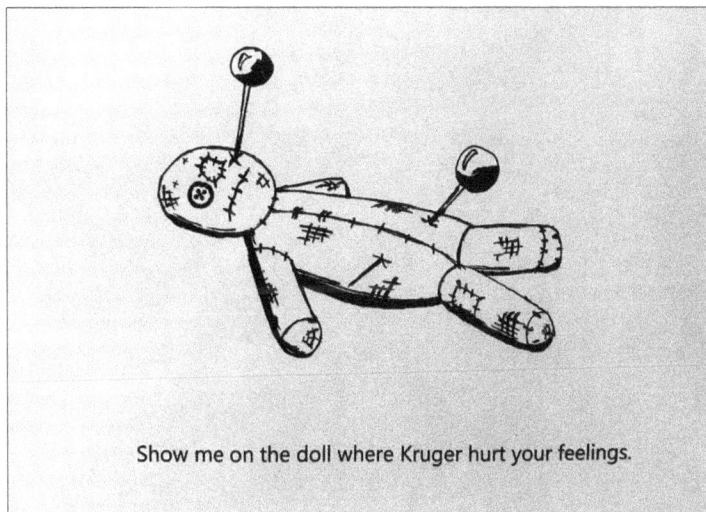

Show me on the doll where Kruger hurt your feelings.

Kandahar 2020. I was at the end of my rope and everyone knew it.

Kandahar 2020. What was I laughing about?

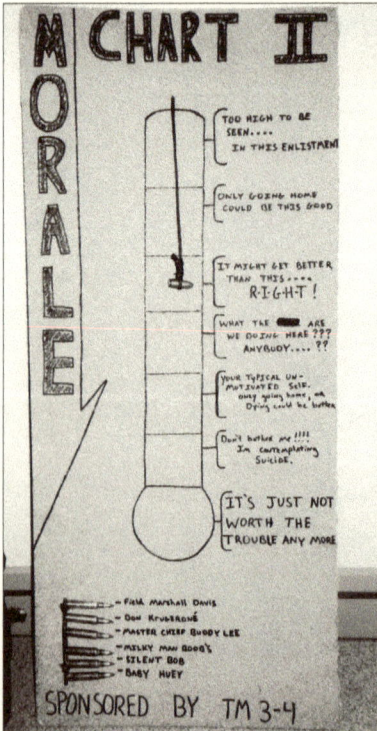

M O R A L E **CHART II**

TOO HIGH TO BE SEEN.... IN THIS ENLISTMENT

ONLY GOING HOME COULD BE THIS GOOD

IT MIGHT GET BETTER THAN THIS.... R·I·G·H·T !

WHAT THE ████ ARE WE DOING HERE ??? ANYBODY.... ??

YOUR TYPICAL UN-MOTIVATED SELF.. only going home, or Dying could be better

Don't bother me !!!! I'm contemplating suicide.

IT'S JUST NOT WORTH THE TROUBLE ANY MORE

→ FIELD MARSHALL DAVIS
→ DON KRUBARONE
→ MASTER CHIEF BUDDY LEE
→ MILKY MAN BOOG'S
→ SILENT BOB
→ BABY HUEY

SPONSORED BY TM 3-4

Morale chart we made in Iraq, 2005

Olympic Center, Seoul, South Korea, 2018

Centennial Warrant Officers Ball,
Oahu, Hawaii, 2018

Battalion Dining Out on the
USS Missouri. Oahu, 2019

Flight line photos taken in negative temps. Fort Drum 2022

Redeployment from Afghanistan, 2016

10-year anniversary. Kailua Beach, Oahu, 2016

CHAPTER TWELVE

J enny's family visited, and we crammed nicely into our small 1,500-square-foot rental house. We had a wonderful week. We all marveled at Charlize's beauty and how she slept through everything. Delusionally, we thought we had one of those babies who slept all the time, and life would be hunky-dory going forward. Jenny's family left, and almost instantaneously, like a light switch going off, Charlize was fussy non-stop and wouldn't sleep. We couldn't figure out what was wrong. She wouldn't sleep in my arms, the crib, or the swing. She hated, and I mean *hated*, the car seat. We learned quickly that the only way she slept was to be attached to Mommy. We were frustrated with her and each other, and we were exhausted.

Still, somehow, we established a routine. Even though I wasn't in IERW yet, I still had to go to PT and do whatever crap they gave us to do during the day to keep us busy. I would get home, make dinner, and do whatever cleanup was necessary from both dinner and the mess the two cats made. Jenny bathed Charlize and tried to get her to sleep for a little while so she could have a break. It was good that we established this routine early on so that when flight school started, we could maintain it. I could get an hour to ninety minutes of studying each night while baths and attempts at sleep were happening. I felt guilty the whole time at SERE

School, knowing that on some of those nights, I was getting a good night's sleep while Jenny wasn't.

Finally starting IERW, the next addition to the routine became buying a bottle, usually Maker's Mark, on Fridays and drinking about half of it on Friday night, the other half on Saturday. Then, on Sundays, we would go to church and after I would study a few extra hours. I would usually take Charlize on Friday and Saturday nights and have my drinks while keeping her swaddled tightly to me so that Jenny could have those two nights to catch up on sleep. I usually fell asleep watching a movie, reclined in my chair, while Charlize was secured to me with one of the baby-carrying wraps we had bought. For the most part it worked, but it was also during these times that Jenny and I would argue about something stupid that would get blown out of proportion by one of us, and then I'd be more pissed off because all I wanted to do was relax from the week.

I have no doubts, as a Christian, that it was more spiritual warfare than anything. It wasn't one-sided. I shouldn't have been drinking at all, and I wish I hadn't started drinking again in the first place.

It all came to a head in August 2010. It was at the point that I hated flight school; Charlize robbed us of any sleep we had hoped ever to see again; and while my drinking hadn't picked up, it needed to cease. Jenny had had enough. She decided to visit her family in North Carolina for two weeks so we could have a break, and her family said they'd help give her the break she needed. I knew it was a test run for her to see if leaving me was feasible. I knew that her family wouldn't be the saving grace she expected them to be, and they would finally see how Charlize wouldn't let us sleep and all we dealt with. I hate to say I was right, but they tried to tell Jenny to "let her cry it out," as if that thought had never crossed our minds.

Charlize did what she did best at that time: cried endlessly until someone came to get her. Jenny got a sheepish apology from her family about their suggestion. Jenny also tried to schedule some outings to get a few hours to herself, but her family reneged on the help they swore to give. Jenny never got the me-time that her family promised her. She hadn't even been there for two days before she called me, feeling frustrated. That was quicker than I expected, but I was no less surprised. My only hope was that this weekend drinker would start to look better in her eyes, and I would get more appreciation for everything I had done. I know I wasn't perfect, but we both had heard plenty of stories about the new mother who has to deal with the children and the husband who walks in the door, sits in his chair, and expects to be handed a beer while he turns on the game, and his favorite dinners are prepared every night. I'm not now, nor have I ever been, that guy. On the contrary, we often hear, "Your husband does dishes, sweeps, mops, and vacuums? My husband would *never* touch those things!" I felt very unappreciated at the time and hoped that this experience with her relatives would bring some appreciation for me. It did, for a while.

After she came back home and we resumed our routine, I tried to help out even more than I had been. One Friday evening, I got home and heard Charlize crying in her crib, as usual. I checked on her and saw she was OK, but something told me to check on Jenny. I walked around every room, which didn't take long in such a small house. I found her in our small walk-in closet, the furthest from Charlize's room. Jenny was crying while sipping on a glass of wine. She looked up and said, "This is the only room I can go and not hear her. She has been at it *ALL. DAY.* And I'm exhausted from listening to it."

I didn't say a word. Instead, I walked back to the kitchen, grabbed the bottle of wine and another glass, walked back to the closet, topped her off, poured myself a glass, and sat there holding her until she stopped crying. After that, I quickly changed into comfortable clothes and grabbed Charlize to wrap her to me. She stopped crying and fell asleep quickly. We ordered some pizza and waited by the door for it to arrive so Charlize would continue sleeping uninterrupted by the doorbell. Jenny took a shower, went to bed, and slept the whole night. A mere eight months into the greatest challenges we had ever faced, both individually and collectively, we were miserable, wondering if it would ever improve.

CHAPTER THIRTEEN

A few days after New Year's, 2011, it was finally time to fly a real bird. I had a new class, a new airframe, and a new stick buddy. There were incoming IPs from the recent PCS season, so my stick buddy and I had a few guest IPs, including our first flights in the Alpha and Lima model Black Hawks. Fine by us. Our first get-to-know-the-airframe flight would be chill. We were to start the aircraft, fly out to a training airfield, perform a series of traffic pattern maneuvers to become familiar with the airframe, switch pilots, conduct more pattern work, return to the airfield, shut down, and fun was to be had by all.

My stick buddy flew first, so I enjoyed the view out the window. I had made peace with the fact that this was my aircraft and put the same effort into learning everything about it that I had put into the TH-67 and flight school in general. We swapped out, and it was my turn to fly.

"Bring it up to hover." Those words made me so nervous! *What is it going to be like?* Like the OH-58 that was so much easier than the TH-67? We had all been out of the cockpit for a few months, so I wasn't sure about my ability either. I slowly started to pull up on the collective while my legs were locked on the pedals, ready to push them in whichever direction was necessary. My right hand death-gripped the cyclic, prepared to move in anticipation of the helicopter sliding in any direction. As the front wheels lifted off the ground, I only made minor corrections to the controls. I only needed minor adjustments as the nose rose higher, and I could tell the tail wheel was about to lift. Finally, according to the radar altimeter, I was at ten feet, and the bird seemed to stay there.

I didn't have to be ready to counter any movement at all. It just hovered. Seemingly on its own. I snickered a bit, and the voice activation picked it up.

The IP asked, "What's so funny?"

"It's so . . . smooth! It just sits here at a hover." I could hear a hint of surprise in my voice.

"What did you expect?" Now she was the one who sounded surprised.

"That's the thing, I didn't know *what* to expect, but it wasn't this!"

We moved out onto a landing pad, took off, flew a pattern, landed to a ten-foot hover, and went for another round. I was in awe of how effortless it all was. The helicopter *wanted* to fly and make the whole experience enjoyable for the pilot. Aside from having a useless heater and no AC, I was beyond impressed.

Advanced Airframe Training, as they called it, was flight school all over again at an accelerated pace, with the aircraft we would fly for the rest of our military careers. The basics common to all aircraft—instrument flight, radio calls, and everything about flying in national airspace—were reinforced, but it was also expected that we were familiar with them. The airframe-specific items, especially Chapter Five's aircraft limitations and Chapter Nine's emergency procedures, were hammered into us daily, and memorizing them was non-negotiable. I used the flip book study cards that students could purchase and memorized them verbatim. It reached the point that I could get through that flip book with my wife quizzing me while flipping through at random, within twenty minutes.

We flew for four weeks, focusing on traffic pattern work, and then had a checkride to assess our proficiency in flying standard traffic patterns and some emergency procedures. We flew four more weeks of Basic Combat Skills (BCS), similar to BWS, but with a Black Hawk this time. Our IP was great as he would have us push our limits with this bird, which was well below what the aircraft could handle. We flew low and fast and developed a comfort level with it that I didn't think many other students had at that time. Lastly, we had a couple more weeks of instruments before we moved on to the night flight. For those of us who chose Mike Models, we would move on to the Mike-specific training before the night flight.

We moved into the Mike Model once Alpha/Lima training was completed. Our Mike Model class only had eight people, four groups of two, and it was the most relaxed time in flight school than any other. Most of the manual for the A/L was also applicable to the M, so it was simply a matter of learning the differences in those models, specifically regarding the 5s and 9s. Essentially, the Mike was going to be a completely different story, as it had all the latest, fully automated flight control systems on board.

The Mike had a glass cockpit. Instead of a dozen gauges on the dashboard and a shared center stack of lights to monitor, we had four iPad-sized screens, two on each side, and a smaller center display with a battery source for extreme emergencies. It also had a flight director for fully automated flight control. Once the flight director was set, all we had to do was turn knobs to stay on course, altitude, or airspeed. The collective and cyclic also had some knobs to make adjustments, which were all displayed on the screens, allowing for fine adjustments as needed during the flight. If all the mechanical systems in the A/L were designed to assist the pilot and offload the workload, the M was designed to reduce it even further.

We began with two weeks in the simulator to learn the subtle differences between aircraft start-up and shutdown procedures, flight controls and screens, and to practice emergency procedures. Then, there were more staging field flights with traffic pattern work. Another thing that made the Mike Model different in flight was drastically larger blades. Autorotation, flying without hydraulic boost (boost off), and other maneuvers exhibited differing flight characteristics compared to its A/L counterpart. A bit more BCS was mixed into this, and then a couple of weeks of instruments before the night flight.

We had a German IP as our instructor for the night flight, as our IP for the last few weeks had been moved up to lead instructor for the Mike Model course. It made no difference; we completed our few weeks of night flying, finished our check ride, which set the tone for every checkride that followed, and were ready to head to our first aviation unit. All my checkrides had gone relatively well. Now, we had a big decision: Where would our next duty station be?

Whenever it was time for PCS, we sent a wish list of duty stations we would like to move to, and if Army Aviation needed someone in my position at one of those locations–winner! You get to go where you wanted to go. If they didn't need you at one of those locations–loser (insert sad face emoji)! They would send you where they needed you. Luckily, we chose Fort Campbell as our top pick, and that's where we went. While I was there for Pathfinder School, I liked the area and thought it would be a good place to be stationed since Fort Campbell was the center of the universe for all things Army Aviation outside of Fort Rucker.

We arrived in late June 2011, and it was a mixed reception for me. We wanted to live on post, but the wait was six months long, so we bought a house because it was cheaper than renting. I arrived to the new unit wearing my ground uniform since I knew I wouldn't fly on day one. I had all my badges on, which also came with mixed reactions. The guys who had done cool things prior, like me, wanted to chat about it. The other assholes said things like, "No one cares about that stuff; wear your flight uniform." It helped me to see who had what type of attitude.

Campbell had recently received Mike Model Hawks, which was good since everyone would be learning together, and some of the usual hazing BS would probably be curbed. All untracked WOs had additional duties, and the first one handed out was "fridge bitch"—with the responsibility of keeping the refrigerator stocked with sodas and snack foods. Every aviation company has a fridge fund that collects money for company events, such as barbecues or family gatherings. The guy who was running it already was OK with it and didn't care about giving it up. I helped him out with it while gaining my duties of unit prevention leader (UPL), (the guy who administers the piss tests), and unit movement officer/air load planner (UMO/APL). With a deployment a year in our future and numerous training events in between, I would be busier as UMO than I was led to believe.

Around mid-August, I had my first flight since arriving to Campbell. One of the staff RLOs (real-life officer, not WO) was an IP and needed some flight time, so I got to fly with him. It was similar to a flight school flight, so I was relieved it wasn't the experience I expected. Most IPs wanted to grind you into the dirt with question-after-question and ask them while you're trying to fly and then finish it with extra homework on things to look up. I had my fair share of those, too, but at least I got to ease into the water instead of being dumped into the deep end. Since I studied a lot of aviation books in my young days of flying when I had those asshole IPs, they weren't as bad as most others have had. Anyway, it was good to get a flight in and see some of the immediate areas that we flew around Campbell. While there wasn't anything beautiful to see, as most of the terrain was flat and unimpressive, the back 40, as we called it, allowed us some decent terrain and low-level flight.

Once I was in progression, it went rather quickly. The first goal was to transition from RL3 to RL2 and then to RL1. These readiness levels tell other pilots, "Hey, he should be proficient in these specific tasks and maneuvers." It ranges from traffic pattern work to night formation flights and flights with sling loads or rappels—stuff hanging from the bird. Once you are RL2, you can fly with MTPs to conduct MTFs (maintenance test flights), and RL1 can fly with any other pilot in command (PC). After making RL1, making PC is a lot of work. There is an extensive checklist of items to mark off on your journey before being let loose with the keys. I focused on RL1, and it didn't take long.

I could focus on flying and work-related matters more easily as Charlize started sleeping through the night, which allowed us both much-needed rest. It took about twenty months, but what a feeling it was to sleep for uninterrupted blocks during the night! We still took turns, but it was usually to be with her and get her back to sleep since she was on solid food and didn't need a bottle anymore. Charlize was such a challenge to us and our marriage so much so that we couldn't fathom going through it all again. We committed to one and done.

Then the light at the end of the tunnel arrived, and at this point we realized that maybe she wouldn't be an only child after all; that one and done commitment went out the window.

CHAPTER FOURTEEN

I was supposed to deploy to Afghanistan at the end of August 2012, but the birth of our second daughter briefly paused those plans. Izzy was born on August 21, 2012, the same day I was supposed to leave. However, I was allowed two weeks of paternity leave, which gave me the time at home to help Jenny settle in and then be on my way. On September 5, after ensuring Jenny had recovered as much as she could—not nearly enough for a woman who had just given birth—I tearfully got on a plane. We all went to the hangar and chatted with people and other families, which was merely a distraction, delaying the inevitable departure. We formed up to walk to the buses and to ride over to the next waiting area of purgatory; I held it all together until we started marching toward the buses.

As my row began to move out, Charlize saw me, stood up, reached her arm in my direction, and said, "Daddy, Daddy!" She wasn't crying, just calling for a hug, or wanting to be picked up; she was a toddler who saw her father walking by and called out to him. It. Fucking. Broke. Me. I'm glad it was nighttime because after we got on the bus I faced the window and cried without feeling like anyone was looking at me. I didn't want to go through that ever again. I told Jenny that on any other deployment they would drop me off outside, we would say our goodbyes in the car, and I would walk away so they could go home.

After a quick stop in Bagram, we arrived at our new location for nine months, the dreaded FOB Shank. When I was in the infantry, I had this delusion that aviation life was better. Well, it wasn't. We had a tiny closet of a room walled off in a tent with eight other people in it. That would be my shoebox for nine months. The food was the same as in previous deployments, which was acceptable for the duration of our stay. The amenities, while lacking from my previous experiences in Iraq, would suffice. We had graduated to the land of smartphones and Wi-Fi, so I was pleasantly surprised to learn that the FOB had Wi-Fi available for personal use anywhere you go. I was grateful to keep in touch with my family effortlessly and wouldn't miss the first nine months of Izzy's life either. It was comical to see the younger guys freak out or get pissed off when the Wi-Fi was down, and I would tell them stories from nearly ten years earlier when the war kicked off. They would still be little bitches about it, but they wouldn't complain around me.

If you know about Shank, you know how miserable it was. And for those who don't, you are going to learn today. FOB Shank was notoriously referred to as Rocket City. We received 110 mm rockets and 80 mm mortars daily. At first, it was laughable. The incoming alarm would go off, and you'd dive to the ground. You'd wait for the explosion, run to a bunker, wait for the all-clear alarm, and then go about your day.

A couple of weeks before I got there, the Taliban tried a coordinated attack on the FOB by exploding 5,000 pounds worth of explosives in a tanker truck right outside the barrier and then having a ground attack to follow and infiltrate the FOB. Unfortunately for the Taliban, they chose a part of the wall that wasn't too close to anything of importance. An infantry unit was training near the area the Taliban decided to breach. Infantry forces made short work of the attack, the engineers patched up the wall, and we were back to business as usual. They didn't try another attack like that while we were there.

The rockets didn't stop. Day. Night. It didn't matter; they didn't stop. They employed a few time delays to launch these harassing fires, so they wouldn't be near the launch sites when the launches occurred, even though they used the same launch sites. One creative way the Taliban used to create a time delay was to freeze the projectile in a block of ice, set it up to launch, and then, when the ice melted, it would launch itself. This would all become discouraging to us when we weren't actively doing anything about it. We weren't authorized to engage if we lost positive identification (PID) on the person or persons initiating the launch. Worse yet, when they set them up for a delayed launch, vectoring where the launch came from would be useless since there would be no one to catch in the act. They knew our Rules of Engagement (ROE) and used them against us, sneaky motherfuckers.

Adding insult to injury—and some people did get injured from these non-stop attacks—was that the only defense we were afforded was to put sandbags around our sleeping areas. We had concrete T-barriers and bunkers to run to afterward, but additional barriers would not be supplied, and we were very exposed in tents. A few of us decided to put some cots in a bunker and sleep in there, but even with a sleeping bag, it got rather cold, and when the alarm went off, it was full of people anyway, so that was pointless.

We were informed that a C-RAM would arrive at the FOB soon. It's one of those Gatling-style guns on the side of battleships that destroys fast-moving, incoming rockets and whatnot. They're cool, but we never saw them in action since it never showed up while we were there. So, we dredged on day-to-day, not knowing but hoping today wouldn't be the day that incoming would be accurate fire.

We had heard the reports of those before us who were hit while walking down the road, in the DFAC, or wherever. I hardened my nerves and acted accordingly when the alarm went off, as I mentioned earlier. Little did I know that this would have a lasting impact on my mental health for years to come.

Flying around Afghanistan and exploring these new areas was quite exciting. We were in RC-East (Regional Command-East) and flew from Bagram in the north to numerous FOBs in the south and as far east as the Pakistan border. Flying nearly every day made time pass quickly.

Thanksgiving Day came, and to boost morale, our BN CO decided to have the companies participate in a Thanksgiving Day Parade. We all made "floats" and towed them around the flight line; some ordered costumes and got dressed up for the occasion, and others, like me, participated because we had to but were ready for the feast that the DFACs always put on. They kept the flights to a minimum so everyone would get a chance to eat and call loved ones. We took a couple of pineapples from the DFAC because they had them, and we wanted them. My buddy and I cut them up and left a considerable plate full in our CP, and to our surprise, everyone ate them. People love pineapple—but not so much on pizza.

We had one other surprise that day. I figured an incoming alarm would go off during the parade, but no. Later, during the feast, none. Talking to my family, no alarm. I went to bed and wasn't interrupted by that damned alarm once. The rockets stopped. *Thank God!*

But we weren't always so lucky. When you're in the military, loss of life becomes a kind of regular occurrence. Don't get me wrong, it still affects me, but with as many deployments as I've had and being surrounded by the possibility of death daily, it becomes commonplace. Still, one such death, a KIA specifically, shook us all a bit in that it happened on Christmas Eve, when a small FOB just north of Shank was attacked and sustained one casualty from an RPG. Some Specialist took the brunt of the explosion and was KIA instantly. Our company was responsible for flying the Fallen Angel mission that afternoon. The rest of us who weren't flying or were on standby for QRF went outside the CP when it was time for them to fly the KIA soldier to Bagram, to be then flown home.

As the Black Hawks transporting the soldier took off, they began to play Taps on all the FOB loudspeakers. At the end of Taps, someone contacted the aircraft on the radio, initiated an internal 3, 2, 1 countdown, and then simultaneously launched flares from both aircraft while everyone on the ground was rendering a salute. It was a touching tribute to our fallen; we flew too many during that deployment. As soon as it was over, everyone standing there saluting didn't even look at one another. We finished our salute, turned, and in unison, walked away in different directions to call or send a message to our loved ones. Somewhat numb to the loss of life, yes. Does it ever get easier for anyone involved? Hell no!

Regardless of the previous incident, Christmas Day was the same as Thanksgiving while deployed, except for the parade we conducted. Flying continued, and I improved significantly. I was recommended for a PC ride, so I prepared for that. I read—or otherwise skimmed—every aviation book on which I could be asked questions. I called the SP that I was going to be flying my PC checkride with Quiz Master Flash, because he would ask some random minutia crap that had no bearing on flying whatsoever but he would just see how deep in the weeds you would study. The deployment was over halfway complete, and I wanted to ensure I got at least 50 hours of PC time before we went home so I could be ready to track when we returned. The PC ride—which is designed to stress you out, overwhelm you, and put you in situations where you have numerous things to deal with at the same time while flying—went well, and I was handed the keys to Black Hawk. The training wheels were off, and I would be responsible for the crew and the mission.

Then, on February 14, 2013, like a sweet gift from a long-lost lover, the rockets started back up. I wish I were making that up—a Thanksgiving Day gift to a Valentine's Day pimp hand slap. There was a "fighting season," and even terrorists need a break to go home, see their families, and refit for the next iteration. They were back in full force, too—numerous times per day, morning, noon, and night. Whoever resupplied them did it well. They never seemed to be out of ammo. Incoming just kept on coming. One of the rules was that we weren't allowed to go to our tents before 4 p.m. since the evening time was when they seemed to hit us less. It was BS.

One day, after a six-hour flight, I was exhausted and wanted to change and relax. At 3:50 p.m., I went to my room. When I got to my room, the incoming alarm went off, and I was livid.

"I don't care if I get hit today. Just end it!" I shouted.

That's when I heard the whistle of the round louder than I ever had. When I realized it was getting louder, I hit the floor and said, "Just kidding. Not today!" *BOOM!*

The rocket hit a T-barrier that was fifteen meters beyond where I was standing and then cowering. Concrete shards rained down on top of the tent. I counted to ten, got up, and was about to head to the bunker when another one went off. I could already hear the whizzing sound, but it wasn't as loud. *BOOM!* This one hit the T-barrier on the other side of the road, thirty meters beyond me. I didn't count; I just got up and headed to the bunker. The all-clear alarm sounded, and I sheepishly returned to my room. It was 4:01 p.m. I didn't bother blowing off that rule again.

In late April, Fort Drum's torch party arrived. Usually, a small group of leaders prepares everything for their deploying party to arrive so it can be as seamless a transition as possible. They hadn't been at Shank for more than 72 hours when an incoming alarm went off. We were in the CP, and everyone got down. *BOOM!* This one sounded closer than most, and when I looked up, our CO SP was looking right at me, and we both said at almost the same time, "That one sounded like it hit the tent area!"

We all went to the bunker and heard the calls for "Medic!" One of our pilots, a prior medic, ran out as more people were pouring into the bunker. One kid came in wearing PTs and no shoes with random splatters of blood and sat in the corner looking a bit dazed.

A PFC from Drum was panicking and saying to her friend, "What are we going to do if it's like this for nine months?"

I felt sorry for her right then, but at the same time, I felt nothing since we had already done the same thing. A few minutes later, we heard that the direct hit had killed some people.

At that same moment, my buddy, the prior medic, said, "That's why I didn't want to be a medic anymore! I didn't want to deal with that. It looked like a haunted house in there."

Two officers didn't even have the chance to get down or out of the tent; it was a direct hit. I couldn't imagine hearing that back home. They lost two people from their unit before the rest departed to join the fight. What a shot to morale that must have been, and even worse for the families to hear that their husbands, sons, brothers, or uncles, whoever they were to whomever, who had just left their home station within a week, wouldn't be returning home.

What's more, their deaths were more than likely preventable if the Army had given us something, anything more, to defend ourselves with besides telling us to use sandbags. More T-barriers, some net system I overheard a mention of, that damn C-RAM they kept talking about, aerial base defense to monitor enemy weapons emplacement, and the ability to take them out in real-time. More sandbags. I know I'm not the only one who felt betrayed. All this combined set me on a long path of mistrust and behavioral health recovery that is still ongoing today.

The relieving unit arrived from Fort Drum, and we were all eager to hand over the reins and get the fuck out of Shank. We moved to transient housing, which was a relief as it was away from where most of the incoming would hit. While in transient tents, we had to monitor radio chatter. One evening, we heard our A-10s flying overhead, calling in and announcing they had PID on some individuals who had just launched rounds into Shank. We all hopped up onto a HESCO barrier to see if we could pinpoint where they were talking about, and sure enough, the area to the south was flat but had a 50-foot hilltop that we could easily see as the sunset was backlighting it. The A-10s were cleared to engage, and we were able to watch them do about five strafing runs each as they reported at least three enemy KIA, much to our cheers and fist pumping in the air. If the little things get you through life, this was one of them.

It was finally our turn to leave FOB Shank. A C-17 showed up as a sweet bird of freedom taking us to Kazakhstan, one of the many staging areas to and from Afghanistan. Rage. Frustration. Confusion. Defeat. I don't know how many emotions ran through me, and most likely many others, as they dropped the ramp. What do we see? A large box, followed by many smaller boxes, labeled C-RAM. No kidding. One of the very guns they had been saying for the last

nine months was on its way to protect from fast-flying pro-
jectiles, and it just came in on the same aircraft we would
leave in. *Would those two Fort Drum officers still be alive if
it had arrived a few months earlier?* I shook the thought off,
got on the bird with the rest of the company still there, and
made our journey home.

CHAPTER FIFTEEN

After four deployments, I wasn't expecting any issues coming home and settling back into a routine. But this time, it was different. Anything that had a tone or hint of that damn incoming alarm set me off. I was on the computer while Charlize was watching a cartoon next to me, some old Strawberry Shortcake show from way back, and there was a tone of some music in there twice that caused me to push back from the desk quickly until I realized I was home and didn't need to do that. After my heart rate returned to normal and the goosebumps disappeared, I tossed that disk in the trash when she was done watching it. Don't come at me; it wasn't one of her favorites, so I knew she wouldn't miss it.

Of course, I was drinking, too. At first, it wasn't much and only on the weekends. Then, it would progress to a drink or two during the week. Eventually, I was drinking most nights, but if I wasn't drunk and making a fool of myself, I was OK with it, and my family seemed fine, too. I kept it together for the longest time, but as some people do, I was drinking to self-medicate. I didn't believe that was the reason since I didn't see anything that I needed medication for. I was drinking to relax at the end of each day, as they were becoming increasingly stressful.

Before we came home, all the PCs had to choose which track they wanted so the BN SP could write recommendation letters, and the unit could set class dates. Most of the tracked aviators had been at Campbell for too long, and a mass exodus was expected, leaving a void to be filled. About a month before we came home, our BN SP sat us all down one by one.

"What do you want to track?" he asked pointedly. He was never one to say more than he needed to, which was never very much.

"I want to track safety," I replied with some hint of reservation. "But I could be interested in maintenance," I continued, assuming we would talk about it some. I was *wrong*!

He said, "Army Aviation doesn't need more tracked safety officers; you'll track maintenance. I'll write up the letter of recommendation with the others. Thanks."

That was that. I should have said safety and left it at that, but I didn't. Maybe I would have enjoyed my aviation experience a bit more if I had tracked safety and not had all the responsibilities of a MTP on me.

I needed a break, so I decided to take a few days off after being home for a couple of weeks and fly to Spokane to see my family. That proved more stressful than I thought, as everyone wanted to see me, and all I wanted to do was be home and get special treatment for a couple of days. I drank the entire time and hid it from everyone. I maintained composure the whole time, keeping my distance and conversations minimal. I wasn't angry; I just didn't want the attention. I appreciated that my family meant well and wanted to celebrate me, but I wanted to escape everyone and everything for a few days. I was able to go out one night with a couple of buddies, and after hopping a few bars, they asked what I wanted to do, and I said, "I want to smoke some weed."

It had been over a decade since I smoked any marijuana, and I thought—I hoped—that maybe I could feel happy. There happened to be a guy my buddy knew at the bar who would have some. If I bought him a drink, he would smoke a bowl with me, so I took him up on the offer. I took two hits, and I was high as a kite for the next four hours or so. Honestly, it felt amazing! I was happy, joyful, and laughing. I went from wanting to hide to wanting to be out and around people. I was tingling from head to toe, feeling warm and fuzzy. I didn't even want another drink; all I wanted was water. My buddies were telling people I was a hero back from war, so people offered to buy me drinks, but I kept saying, "I'm good," and enjoyed water after water. I wondered how I could get this feeling on the regular, but the following day, I panicked about my career and coming up positive on a drug test, so I let it be. I enjoyed my relief for a night, and it was back to home and reality.

Reality bites.

"I was in an accident, and I think the car is totaled," my wife's voice trembled on the phone. "I'm in an ambulance, and they are taking me to the hospital."

"Are you OK?!" I asked in a somewhat panicked tone.

"I'm OK, but my neck hurts, probably from whiplash, and they are taking me in for X-rays," she explained.

She was OK, and I include this story not to highlight that it was time for a new car after she totaled her 2004 Mazda 3 that she'd bought brand new nearly ten years prior. I bring it up as another example of fantastic leadership in the form of a company commander.

As soon as I informed our CO that my wife had been in a car accident, she didn't even hesitate to say, "Kruger won't be going to the field with us in four days; what do we need to do to fill the gap?"

I didn't have to argue, fight, or otherwise convince anyone that I needed to be with my wife. She knew it was the right thing to do and let me skip the training requirement without a second thought. If only all commanders at all levels viewed this similarly, the military would likely have fewer retention issues.

I was scheduled to attend the Maintenance Test Pilot Course (MTPC) in October 2013, but a government shutdown caused a change in plans. I would attend the Aviation Maintenance Officers Course (AMOC) in January 2014 and MTPC following that in April. I went with one of my coworkers, which made the process more enjoyable as I didn't have to learn all new people, and we had each other to study with. Plus, our wives were good friends with our kids around the same age, so they had support back home from other friends, too.

AMOC turned into a battle against boredom. Daily classroom attendance being released between 2 p.m. to 3:00 p.m. Every test was open book, so there was no need to study. I would hit the gym, grab dinner at the DFAC if we weren't going out to eat, and then drink a few before heading to bed. Wash, rinse, repeat. The Seahawks destroyed the Broncos in the Super Bowl that year, and it was fantastic to watch. I watched it with a buddy there for MTPC who happened to be a Broncos fan. Everyone was talking shit about "defenses don't win Super Bowls," my buddy included, so it was fun to see them win in such an undisputable fashion.

Back to AMOC. They teach everything you need to know about the bookkeeping side of maintenance, including higher levels of maintenance, and how to effectively run a maintenance program. If you were like me and weren't prior aviation and had no plans of tracking maintenance until a couple of months prior, then it was all foreign language. It wouldn't be helpful information to me, and in the end, most of us left a critique that for MTPs, it should be a course we attend after being in maintenance for at least a year.

AMOC had some unintended consequences for me. The first thing my buddy and I did when we arrived was hit the

Class VI. That's a liquor store for non-military types. After an honest attempt to cut back my drinking drastically and somewhat succeeding, I decided it best to buy a few larger bottles of things to have around so that I could graze at my leisure while in a room all by myself for seven weeks. A huge misstep on my part.

As one might suspect, it didn't take long for that to unravel. "It's only a couple of drinks," I would say to myself on a random Wednesday with nothing to watch on TV, after a workout, after dinner, and no one to hang out with who wasn't drinking anyway. I convinced myself that as long as I didn't have a hangover and could still do all the daytime functions. I was in control; I was good to go.

We finished up AMOC and headed back home to Campbell. I decided to quit drinking again to help in my endeavor to overcome all my struggles, which worked for a while. We had one month, and it was back down to Rucker for MTPC. Based on my buddy's recommendation, who had been there previously, we focused on 5s and 9s, as MTPC was a gentlemen's course, and we weren't expected to know anything that hadn't been previously taught to us. This time, my sobriety would remain intact. I had planned to go to the flight line with my buddy, attend the classes, and once that was complete, get a workout in, have dinner, study for an hour or so, call my wife, shower, and go to bed—every day. On weekends, I would sleep in and do whatever anyone else did, and everything would be peachy. Surprisingly, it worked out that way.

MTPC was an excellent course. We signed into the course and received the publications and regulations we needed. Since everyone was using tablets, we only needed a MTF manual—more aptly called a checklist—even if it was a manual. We were to highlight, color, and memorize the briefing associated with the Big Five checks. Such checks required the MTP to manipulate power control levers while the PI needed to maintain or adjust the flight controls during the check. But first, we needed to learn how to start the aircraft, so it was back to the simulator. *But Chris, you're already a PC, don't you know how to start the helicopter?* Yes, but now there was a particular way to start the aircraft while looking for any maintenance malfunctions, whether electrical or mechanical. The first few days were at the simulator, running through these motions and explaining to an instructor what we were looking and listening for. Exciting stuff—that's sarcasm, folks.

Next, we moved to the aircraft and did the same on the flight line. First, I performed all procedures up to engine start, then shut it down. My buddy hopped up front, did the same thing, and that was it for the day. Throughout this time, we had classroom lessons that delved deeply into each component of the Mike Model Black Hawk. The classroom learning was mostly in-depth and top-notch. There were cutouts of the entire engine and transmission assemblies to see what was happening inside them while in operation. We had all the classes on PowerPoint, so it was easy to study later and see the entire breakdown of each component, along with any necessary maintenance information associated with it, such as pressure, revolutions-per-minute rate, or other relevant details. I always had a copy of these classes on my iPad as my first reference when we had an issue I was unfamiliar with. The PowerPoint classes generally pointed me in the right direction. We repeated the above process for all the required ground, hover, and inflight checks, starting with the sim and then in the aircraft

June 4, 2014, was checkride day. I was ready to go home. The sooner we knocked it out, the sooner we could hit the road! My maintenance examiner (ME) and I made our way to the aircraft, and during the preflight, he asked numerous questions. Of course, they were those next-level questions since I was a MTP. He "chased the rabbit to the hole" by leading me in the right direction for the answer, and then it would come to me, so I did fine on the oral knowledge evaluation. We hopped in the aircraft and went through a quick runup since it was already getting hot on the flight line, and we wanted Big Windy blowing.

"We'll do the Big Five," he said. "When we get to the inflight checks, we'll do the autorotation check with both PCLs (Power Control Lever) to IDLE and the engine checks on one engine only. If you're good to go, we're done. If not, we'll do it on the other engine and back and forth until you *are* good. Questions?"

"Nope," I replied.

"Awesome! Brief me on the tail rotor servo transfer check on our way to the sod so we can get right to it."

We moved over to the sod, knocked out the tail rotor servo transfer check, and continued. The generator under-frequency check was next, and I was good to go.

"Calling tower, set up the flight director to 5,000 feet, 80 knots; we'll use that as check altitude on our way to 6,000 feet; give me your brief on the climb out."

All these maneuvers required a MALE brief: Maneuver, Abort criteria, Limitations, and Emergency actions. I gave him my brief for the auto, and we climbed to 6,000 feet. One PCL to IDLE, initiate auto, and the second PCL to IDLE. We fell faster than a bird hit by lightning. At five thousand feet the rotor speed checked out, I quickly moved both PCLs to FLY, on to the next check. I finished the max power and VH (Highest velocity is the fastest the aircraft will fly in the current configuration. It verifies that flight controls are rigged properly.) checks with no issues. We returned to the parking pad to shut down the aircraft and head inside, knowing there wouldn't be a graduation ceremony. The ME handed me my graduation certificate and the paperwork to confirm that I had completed the course and was a MTP.

He gave me a handshake and said, "If you have issues you can't figure out, call us. We love to help!"

It was BS. I called a time or two, but no one called back to help. Either way, my buddy and I left the hotel and headed home. Newly minted MTPs, and we were ready for anything the aircraft threw our way. Hysterically, we weren't prepared for anything more than a complete MTF.

CHAPTER SIXTEEN

Back at Campbell and it was finally time to get the hernia surgery that I had been putting off for well over a year at this point. It was an inguinal hernia on my left side that, other than causing something the size of my pinky to protrude from my lower left abdomen, didn't bother me at all.

Our company was slated to attend JRTC, which I did not want to do since I had just returned from two courses. It's hot and nasty down there that time of year, and I had already been gone for four months anyway. Also, it was time for the old snip, so I had them do a two-for-one special while they were already down there. It was worth the six-week down slip.

A few weeks later, while I was still recovering from surgery, Jenny and I headed down to Nashville for the day to take the girls to a splash pad, a park, and walk outside the downtown area. I was driving when my phone rang, but it didn't have a name with the number, so I left it to voicemail.

Jenny took a listen and said, "It's from Flip!"

Flip was an infantry buddy of mine around the time I was preparing to transition to aviation. He was shadowing me to take over my position as Air OPS NCO.

Jenny got relatively quiet, then she turned to me and said, "Sammy was KIA in Afghanistan yesterday."

You could have heard a pin drop over the road and engine noise. We were both stunned by the information, devastated. I was driving, but I wasn't at the wheel. It seemed like all thought had stopped, and I was running on autopilot.

"Are you OK?" Jenny asked.

I blinked a couple of times, snapping back to reality. "Yeah."

"Do you need to pull over or stop?"

Still not all there mentally, "No, I'm good."

I was numb. I thought a million things and nothing all at the same time. It's like hearing your brother just died. I don't have a brother, but he was as close to one as I could imagine. When someone tells you, in so many words, that you were their mentor, that's a brother! When we were deployed in 2008, he had a spicy tuna wrap with bacon at the DFAC. He ate one daily, and I made fun of him for it.

"Have you ever tried it?" he said with a hint of sarcasm in a curious tone.

"No."

"Then how do you know if it's good or not?" With that, I tried it, and damn, he was right; it was fantastic! His mom was Filipino, so we talked about lumpia, chicken adobo, and other Filipino foods we ate as family holiday traditions.

I knew as much as a guy could from rolling around Iraq with him and his team, and I know one thing for sure: he deserved better than getting shot in the neck and dying in the Middle East. I can rarely tell a story about Sammy without tearing up. When we got home that day, I told Jenny I wanted a drink, and she understood, though she probably knew it wouldn't be a good idea. I had a glass of wine since we had a few bottles on the shelf, then one turned to two, and I stopped after three.

Sammy's funeral was in Fort Walton Beach, Florida, and I was not going to miss it. At least thirty guys I had known from various years of my time in the LRS unit attended his funeral. It was turned into a reunion and a wonderful one at that. I had never been good at keeping in touch with some of those guys until then. If anything good came from his funeral, it was the opportunity to reunite with other brothers. I don't use that term lightly; they are all exactly that. We met at the viewing, which I couldn't do. I went up and saw him in the casket, and it was all I could do not to break right then and there. His wife came up and talked to us, asking how we knew Sammy, and I kept it brief while giving as much detail as I could, despite losing my breath and having a tight throat. After that, I saw a couple of other familiar faces and went to talk to them, but mainly so I didn't have to see my fallen comrade lying there anymore.

We went to get food and drinks, though I only had one drink to commemorate with everyone and then hung out in a quieter section with other guys with families too. It was nice to catch up with guys I hadn't seen in six to ten years as we had all seemed to pick up right where we left off. No awkward conversations, just happy to see each other and catch up. Then, we put on shirts someone had made honoring Sammy and went outside to get a picture together. Some of the guys in that photo are no longer alive today either. That night, one of the guys we all worked with had a get-together at his house, so half of the group gathered there and continued to drink and reminisce. Again, I kept it easy with alcohol.

The funeral came, and the speeches flowed. Five of us told stories about Sammy, and I wanted to tell the story I said earlier, but I couldn't get that lump out of my throat and knew I'd sound like a bawling idiot, so I didn't. At least it's in here, so I still get to share it. It was a sight to see when leaving the funeral home for the cemetery. The entire community came out, waving flags, with some veterans in uniform showing their support. They lined the road the whole way to the graveyard. It was so awe-inspiring to see a community line up to honor one of their own who had fallen in the war. I teared up then and am tearing up as I write it. The University of Houston football team also wore his number on their jerseys for the next season in his honor.

I don't know if it was a combo of all the previous month's changes and stressors that added up, but besides my flying occupation, surgery, and the loss of a friend, we had to leave for more training, and I was irritated. I was barely finished with my hernia recovery, and we were on our way to Alamogordo, New Mexico, for more HAMETS training. At first, I was irritated as it was another month-long training event, but BN was a bit smarter and decided to rotate the companies out so that only a few key personnel would be there the entire time. It was rather infuriating that we knew we would be deploying to Afghanistan again in seven months, so everyone wanted to make the most of their time at home with their families. Luckily, the chain of command thought the same way, and that made for a reduced time away. An agreeable chain of command doesn't happen often, so I counted my blessings that I was only out for a week.

The training was identical to the one we conducted previously at Fort Carson. If I remember correctly, the highest altitude we flew in New Mexico was only 6,500 feet. My buddies and I "borrowed" a van on our last day, drove to a fire lookout in Ruidoso, and hiked to the top where a marker was 9,641 feet—some HAMETS training. We went and hiked higher than we had even flown!

These are the final details that bring us full circle to where we started at the beginning of my story. I had allowed all of life's situations and obstacles to dictate my feelings, emotions, words, and actions. I used drinking as a coping mechanism, and with Air Assault School having half days sometimes, it allowed me more time to drink. All these factors combined to boil over when being pushed and I took it out on my wife.

CHAPTER SEVENTEEN

Remember the story in the prologue? Well, this is what happened next.

On that afternoon in October 2014, when the broken parts of me I'd tried to hold together shattered, Jenny had every right to call the police, regardless of my trying to stop her. I should have been hauled away in front of my kids and the neighbors for what I did. But for whatever reason, she agreed not to and said, "Fine, I'm calling the pastor!"

"Call whoever you want." I got in my truck and left.

I went down the road a mile, pulled into a parking lot somewhere off Tiny Town Road, and just started bawling. I knew the National Crisis Hotline number, or whatever they were called, dialed them up, and began the process of getting the mental healthcare I needed. Whatever was going to happen next, I knew I needed help getting through it.

"Are you thinking of hurting yourself or someone else?" they asked.

"Yes."

"Have you tried to hurt yourself or someone else? Please remember that we must inform the law authorities if you have committed a crime."

"No." I lied, though I justified it to myself since I didn't hit, kick, or otherwise break anything on my wife. After speaking with them for a while, they set me up with free counseling, and I knew that would be beneficial in the coming months. I didn't feel any better about anything I had done, nor should I have, but I took a step.

I didn't know what to do, so I called my parents. I informed them of what I had done and that I would most likely be going through a divorce. They sat on the other side of the line in astonished silence.

"I don't need you to say anything encouraging or condemning. I feel horrible as it is already," I began to ramble. "I can't even look at myself in the mirror right now." My eyes welled up with shame over my actions. The silence on the line allowed me a moment to ponder how I had arrived at my current state, both physically and metaphorically. "I just wanted you to hear it from me firsthand."

"We're here if you need to talk, and we'll be praying for you all in this time." That's all I needed from them, and then the phone rang. It was the pastor.

"I've asked your wife to come and meet me in my office, and I want you to come. Understandably, she is too scared to see you on her own."

I agreed, and though it was only five minutes down the road, it felt like an hour's drive. I figured the cops would be waiting for me, and it was all over. People would know Kruger lost it and had to turn in his wings. I'd be out, divorced, figuring out my next move to pay child support, and probably going back to Washington to live with my family so I could afford to support my ex and kids. Would they cuff me in front of the church, no less? I'd been there before, just not at a church. My mind raced with everything I deserved to have happened to me. I arrived, and not a cop car was in sight, just four vehicles, mine and my wife's included.

I walked in and up to his office where he was sitting at his desk, my wife in a chair opposite, and one open chair away from her. "Have a seat," the pastor said.

There is a lot, so I'll summarize some key points. The pastor said that he almost called the cops himself but didn't due to my wife's urging.

"Jenny doesn't feel safe around you and doesn't want you at home. Since you aren't going to jail today, you need to figure out somewhere to stay while whatever happens next happens." I agreed. We all talked briefly about the next steps for both of us, and we decided that Jenny should leave first to get home and have time to get a friend over while I figured out where I would stay. Then, I would go and get the clothes and things I needed from the house and leave until Jenny was ready to talk or take the next steps. Again, I agreed to this as I was in no position to argue. I would have rather shriveled up into nothingness than live with myself and my thoughts and remember what I had done to the woman I love in front of our children. I stayed with a coworker's family, who lived a few minutes away from our house, in a spare room they had while I waited for Jenny to decide the fate of our family and I finished the stupid Air Assault course. I decided that I should also take a drinking hiatus, of course, and committed to sobriety once again.

After a week or so, my wife called me over to the house to talk. She ensured our neighbor was home but didn't ask her to come over as she would trust that I wouldn't lay hands on her again. She also sent our kids to the neighbor's so we could talk uninterrupted. I can say that I have never done that to my wife since, but I am also still so regretful that I ever did it in the first place. Even as I'm typing this story, I feel so much hatred for the person I was. I may not have forgiven myself for it; I don't even think I deserve forgiveness from anyone. How could a husband do that to a wife? How could I do that to my wife?

We talked about everything for hours. What we'd each been feeling about the other, about life, about the past—our past—both individual and together.

"I'm stressed because all the pressure is on me, and I didn't want to stay in the military, but you didn't want to teach anymore, so *we* agreed that *I* had to stay in the military . . ." I began.

"Having kids hasn't turned out to be what I expected," she interjected, and we talked on each subject together, diving deeper into the struggles of these topics that we had only scratched the surface on before.

"I feel like a failure as husband, father, and pilot because of . . ." I stated, listing off numerous reasons that are why I coped with alcohol.

"I feel like a failure as a wife and a mother because of . . ." she continued, listing reasons similarly.

We discussed the future and our desires for it, individually and as a family. We took the time to understand one another's burdens more than we had in a long time, maybe ever. We started our conversation in the early afternoon and by the time we got done, it was dark outside, even though it was October and it began to get dark earlier. Though it was a good and much-needed talk, we decided I should stay away for a few more days so we could reflect individually.

Some much-needed discussion and healing came from it all, and in the end, my wife forgave me. The hurt was still there, and so was the regret, but God can use all situations for good to those who believe, and I believe that's just what He did.

CHAPTER EIGHTEEN

I finished the uber-annoying Air Assault School, and a coworker pinned on my wings. I returned to my buddy's house, packed up my stuff, and went home. It was mid-October 2014, and we were deploying in just six months, so I wanted to spend as much time at home with my family as possible, recover from recent events and prepare for the time apart ahead of us. I knew that a looming deployment also meant Jenny would internally ramp up the panic, leading to a few arguments. I prepared myself for that, but with the holidays ahead, I focused on making that the best time possible. We have had the "if I don't return" chat too many times, and I knew we would be having that again soon, but for now, I wanted to put all that aside and just be with my ladies.

In the first week of January 2015, our company had to go to JRTC. *What the fuck! We already went. Why are we going again?* Although it was a waste of time, it was a valuable learning experience for me, and we didn't have to play the usual games since we never entered "the box." We could self-deploy and return to Campbell immediately upon completion, so we were only gone for about eighteen days. I gained hands-on experience troubleshooting aircraft problems in the field, which proved invaluable during the deployment. Additionally, I had the opportunity to work with the entire crew we would deploy with.

The months leading up to the deployment flew by. Before I knew it, it was April 2015, and my girls and I were headed to the airfield to see me off. These were always quiet rides. No radio, the girls were making noises, playing as kids do during car rides with their siblings, and I was blankly staring out the window. *Here I am again. Here we are again. How many more times?* I was informed that I was supposed to PCS after MTPC, but the unit needed MTPs for the deployment, so my orders were pulled. I could've been going somewhere that wasn't deploying at all or somewhere that deployed sooner. I was growing tired of people making decisions on my behalf and determining my future according to their needs, rather than my own wants.

We arrived at the hangar, and I asked the girls to stay in the car while I took my bags inside. They obliged. After dropping my gear and checking in, I headed back to my ladies. I made my way around the car, hugging and kissing each of them. I walked over to the driver's side, gave my wife a long hug and kiss, wiped a tear or two from her eyes, and just held her as she held me.

Finally, I said, "It's time to go." I gave her one last kiss, popped my head in a side window one more time for the girls as Jenny was getting in and buckled. After they were all ready to go, told them, "I love you and will talk to you as soon as I can." As I turned to walk away, I heard "Love you, Daddy!" almost in unison. It was never easy leaving. It only got more challenging over the years.

From Campbell to Kuwait, and then Bagram for the usual in-briefings, then Jalalabad—which will be referred to as "J-bad" from here on out. Once the time to leave comes, everyone wants the same thing: to get to where we're going and get to work. No one ever wants it to drag out, and the sooner we get to work, the faster time passes, and the quicker we get to go home. Most deployments begin the same way:

the first three months pass quickly, and then time seems to stand still. For the next three months, every day seemed like three; they didn't go by quickly enough! Then we got into the home stretch, and the last three months seemed to pass quickly, especially when the relieving unit arrived.

However, for now, let's go back to when I started this adventure. I learned shortly before deploying that I would be the lone company MTP for the duration. Two other MTPs were available to help, but they became unreliable to me due to other duty obligations. I was expected to be more proficient than I was, and this deployment would be the same as the rest of my military career: a trial by fire with little to no mentorship. I had to rely on everything and everyone I could to succeed and make the TF successful. We had eight aircraft, of which two would be located at other FOBs, and one would be undergoing phase maintenance at all times, which meant that we would have five aircraft available daily at Jbad. Sounds pretty good, right? It did until our TF wanted to fly six aircraft worth of daily missions! Begin the balancing act.

Let's look at the scoreboard quickly. Eight aircraft, three locations, one permanently unavailable, and a TF trying to fly more missions than we have aircraft available. Three MTPs, of which I was the only one constantly available to fly MTFs. I was expected to run and fly maintenance while being a PC, flight lead (FL), and AMC. Occasionally, I would get to be a PI. Manage all that for nine months, and any sane person's head would explode. After a couple of months, any sane person would be *begging* for help. I never did because I knew it would never arrive.

Additionally, even though the Kiowas had been removed from the Army Aviation inventory, I was even more frustrated that I never got them as my pick. Remember that guy I mentioned who was first pick in our airframe selection?

222 • CHRIS KRUGER

He was in Afghanistan instructing the Afghani pilots how to fly the MD-530s. He told me command approval was all I needed to go up on a flight with him. I jumped at the opportunity but was informed our command was saving those flights for our enlisted personnel planning to go to flight school. It was an incentive flight for them, as it were. A bunch of the other pilots sure did get to take a flight, though. Of course, they were W4s. Forget about the hardest-working guy in the TF. Whenever someone asks, "Aren't you glad you didn't fly Kiowas since the Army got rid of them?" I tell them that story with an emphatic, "NO!"

A few weeks later, we were in the full swing of things. J-bad was a small FOB, and I loved that fact. Aside from the flight line, we only had a short walk to the DFAC, laundry, shopette, and barber. We had a complete gym in one of the clamshell hangars and everything else we needed in the immediate vicinity. The small footprint helped us manage all our maintenance needs, so we didn't have to go far or across a FOB for food or anything else. The only other saving grace is that we weren't running twenty-four-hour ops either. We had morning and afternoon flights. The morning flight was briefed at 5 a.m. to take off by 7:30-8:00 a.m., and the afternoon flight was briefed at 11 a.m. to take off by 1 or 2 p.m. Any maintenance that couldn't be completed by 11 p.m. was handed over to the civilians to complete, but this didn't always mean it was ready in the morning. We could only hand the civilians stuff that was in dire need of being completed as they had one negative quality about them: they were slow. They did good work but at the cost of speed—another factor in the balancing act.

I would say that this is about the most fun I've had on a deployment, despite all the challenges. When we flew, we would have the best time as we could. We flew low through all the canyons and ravines we could find. Anytime I had infantry guys in the back, my goal was simple: lots of hollering and trying to make someone puke. One day, after dropping a couple of squads off for a village engagement, our FL wanted to fly NOE through our test fire area, which was just a dried-up riverbed. As the AMC, I told him, "Sure," and to announce any sharp left turns he was going to make before making them so I could anticipate it as I was going to fly one disk separation on his left side. It was midday, and we were low on fuel, with no personnel on board, the doors open, and at an altitude of 1,600 feet. We had substantial power margins, so I wasn't worried about anything. He brought the aircraft to twenty feet off the desert floor. I matched. I was at one disk, closed it to about half a disk separation, and locked it in. He turned right, I pulled in a little power, turned right, and stayed half a disk. He turned left, I reduced a little power, turned left, locked in, half disk. If he came up or down over a terrain feature, of which there weren't many, half a disk, locked in position. A crew chief thought it looked badass, so he pulled out his phone and started recording.

We returned to Jbad, refueled, shut down, grabbed some lunch, headed back out, picked up the grunts, and took them back to J-bad after buzzing some hills and low ground. Mission complete. That's when our BN SP approached me. "Why the hell are you out there flying at five feet?"

"What are you talking about?" I said, thinking he was exaggerating.

"I saw the video of you guys fucking around in the test fire area. What made you think that was a good idea?"

I defended my position by reiterating everything I had mentioned earlier: daytime, ample power, a well-known area we flew through daily, experienced crews, and so on. He mentioned that the PC of the lead aircraft, though a chief warrant officer 3 (CW3), had less experience and hours than I, and I should've considered that.

"That's why he was lead," I said, "so I could keep an eye on him."

Ultimately, he said that my PC orders almost got pulled, but the TF needed me to keep flying. That reminded me of my Article 15 years prior. It's always nice to be needed. I found out later that our company safety officer was the one who defended me in the whole ordeal. He showed the BN SP, BN Safety, and BN CO my risk assessment, which showed no restriction to flight altitude and that I was legally flying what I had briefed. That chat from the BN SP was a scare tactic since that was all they had to use. I was grateful that someone had my back since I was working my backside off anyway!

Still, the crew chief who made the video violated the first rule of deployment video making: never post to Facebook. I tell everyone I fly with to take all the pictures and videos you want, but don't post them to social media! But he didn't just post it on social media—much worse! He posted it to Facebook, *and* the BN SP was one of his Facebook friends because, as the crew chief stated, "He has a hot daughter that I want to friend request so I can try to get together with her." This deployment surprised me as it wasn't the fuckin' LTs that made it challenging; it was the lower enlisted.

I did some rough mental math on this after the deployment. I know I flew some five-hundred hours, but the maintenance breakdown was insane. I was flying a pre-or-post phase test flight every week, which included all the main and tail rotor blade and high-speed engine shaft balancing, engine flushes, and any other MTP-specific maintenance and troubleshooting that popped up.

It didn't help that Jbad was down in a river valley, hot and nasty until almost November. One guy called it "J-Africa Hot!" No lie, it was still the hottest I've ever been, Iraq included. It made all the aircraft run-ups and shutdowns even more miserable, covered in sweat. In addition to all the missions we flew all over the eastern area of Afghanistan, I was beaten down before the halfway point.

The last significant event during this deployment was a C-130 crash during takeoff. I had just returned from my PM flight and was about to fall asleep when I heard the C-130 powering up for takeoff. From my hut, I could hear them picking up speed down the runway, the liftoff, a strange change in engine noise, and then a faint thud and quieted explosion. I thought, *Nah, I'm hearing things wrong,* and pushed it out of my mind. A minute later, there was a knock at my door.

"Hey Kruger, I'm coming over from the TOC. Did you hear the explosion?" one of the LTs on duty asked solemnly.

"Yessir, what does that have to do with me?"

"A C-130 just crashed on takeoff. Your crew has been given a two-hour duty extension," he stated rather reservedly. He must have known it was ridiculous, but I know it wasn't his call.

"A two-hour duty day extension, for what?" I begin to blast but caught myself and calmed down a little. "A flight to Bagram? There and back is over two hours!" I said.

"I don't know, probably nothing. They just want you standing by."

I got dressed, got the rest of the crews up, and twenty minutes later we were told to stand down as firefighters had been sent with the infantry guys for security to put out the fire. The rest would be dealt with in the morning. Ten people perished in that crash. One of the engines failed on takeoff, causing the aircraft to roll to the right and subsequently impact the ground just outside the FOB wall. What's crazy was that an infantry unit about to head home was supposed to be on that flight. Instead, a large container was loaded onto the aircraft, so the company couldn't board the C-130. The CO decided to wait a day until his entire company could get on a bird and fly out together. What was even crazier was that if the C-130 had rolled left on takeoff, it would've crashed into the FARP, resulting in an explosion and gas fire that could've spread to our living quarters and resulted in the loss of numerous personnel and possibly all our helicopters that were in the vicinity as well.

It is unbelievable to think how many times I've been close to death during my military career. I have no doubts that God has been looking out for me.

Again, we flew another Fallen Angel mission. It took four aircraft to take all the fallen to Bagram to begin their journey home. As the missions never stopped, the PM flight took the fallen to Bagram. Seeing them take off, disappear into the night sky, and execute a couple of left-hand turns, while launching flares in unison from pitch black, was awe-inspiring. We could hear the countdown over the radio, and all four launched them in perfect, simultaneous harmony. You could momentarily make out the outline of four Black Hawks through the flares, then disappear. Each flare launch tosses out two visible flares. To see it twice was a tear-jerking moment of reflection. The crash didn't happen around any major holidays, and it could have been a lot worse with thirty more people on board, all set to begin the trek back to their families and loved ones. Still, the unnecessary loss of ten people for a war that a later administration would abandon seems even more senseless than it did at the time.

Jbad had one benefit throughout the nasty heat and what looked like cancer-inducing smog on a regular basis. We only had two rocket attacks the whole time we were there. I was grateful beyond measure for that. The two we had didn't induce the type of PTSD reaction I feared that they might have from prior experiences. The incoming alarm went off, and since it wasn't so much a battle drill as it was at Shank, we all looked at each other confused, like, "Is that *really* going off?" Then, we all just kind of shook it off and asked each other where the bunker was and if we should bother going, which we didn't. The all-clear alarm rang within a few minutes, and we didn't overthink it. The second time it went off, we didn't bat an eye.

In all its infinite wisdom and glory, the Army decided to hose two units on Christmas and New Year's. I was optimistic about being home for one of those holidays, as I had been home two weeks before the nine-, twelve-, or fifteen-month mark on every deployment I had been on. Not this time. Not only would we miss both holidays, but Fort Carson, the relieving unit, was sent a bit early, so all their personnel would be missing out, too! Their last group of folks left on Christmas Eve, which, in my opinion, was pretty messed up. We handed over operational authority to Carson on January 1, 2016. On New Year's Day, we started a *Band of Brothers* marathon for anyone who wanted to watch. Since I hadn't seen it before, I was hooked. Not much has changed in wars in nearly a hundred years. I think that if politics and shitty leaders were removed from the scenario, wars would be over a lot quicker.

Finally, on January 11, 2016, we arrived home at Fort Campbell. It was cold, but my wife wore this beautiful red dress and my daughters also looked lovely in their dresses. I was relieved to be home and ready for another month of leave. We planned to head to the mountains and stay in Gatlinburg, Tennessee for a week. It's always nice to have time to talk with my wife and plan for the upcoming months without being surrounded by huge distractions. I would leave in March for the Advanced Warrant Officer Course (AWOAC) at Fort Rucker, while we prepared to PCS to our next duty station, Schofield Barracks, on Oahu, Hawaii. We

228 • CHRIS KRUGER

were ecstatic about this move, as my wife's background is primarily Pacific Islander, with a strong Filipino influence.

Before we PCSed, I went to visit my family in Washington for a few days by myself. I spent most of my time secretly drinking to make it more bearable for myself to want to talk to all the family that came to visit. It did help to calm my nerves, and I was keeping it to a minimum, but I knew I was only kidding myself: I was turning into a drunk again.

One night, I didn't want to drink, and since weed had become legal in Washington state, I figured I'd give it a try. I thought it was somewhat of a novelty that I could buy weed right next to the place where I'd taken Taekwondo lessons twenty-five years prior. I didn't know how the whole process worked, so I told the dude I wanted something to relax, not too strong—a nice, body high type of buzz. He gave me something called Snoop's Dream, which turned into a nightmare. I smoked a bowl near the house, got home as everyone was in bed, and thought, for old times' sake, I'd play some video games on the Super Nintendo my parents still had on the TV in the basement. I quickly lost interest in that and decided to go to bed. I fell asleep fast but woke up an hour later in a panic as I started running scenarios of failed drug tests and getting kicked out of the Army through my head. I decided to toss the rest of it, and after I got home, I told my wife about it. She said I was stupid, and we put it behind us, and I haven't wanted that in my life since.

AWOAC is another useless thing I have wasted months on. Another one of those things that some W4 had said, "You need to have this completed to be competitive for promotion." I'll sum up AWOAC like this: I was a miserable drunk through the whole thing, was in the top 20 percent, and made it on the Commandant's List. During my time there, I wrote a paper on the UH-1H Huey as the workhorse of Vietnam, which I am proud of, and which probably inspired me to write more. I'm grateful that AWOAC opened my eyes to the possibility of doing something beyond what I thought I could do in the first place.

On June 10, 2016, we landed in Oahu, Hawaii. With a crunched timetable, moving house early, a car even earlier, and one cat to transport, we considered ourselves fortunate to have so many friends and family help us along the way. I had signed into the unit, and since they had a huge island and military branch-wide training event coming up called Rim of the Pacific (RIMPAC), they decided to make me wait until that was all over to begin my progression flights for the island, as well as learning the ins-and-outs of medevac. I showed up to work, and since I couldn't fly and didn't have a house, they sent me on my way to "spend time with my family and enjoy the island," which we did.

In mid-July, we moved into our house on Schofield, began my flight refresher training, and experienced flying to the other islands. I already liked the change that medevac would bring to my flying experience. Little did I know, I would fly two real-life medevac missions on the islands, one saving a Marine's life. It was beautiful flying around Oahu and seeing some places my family and I had already driven to and visited. The training area had a few waterfalls accessible only by long hikes, so it was fantastic to fly right up to them and get some pictures.

After finally settling and completing another Annual Proficiency and Readiness Test (APART), I learned the ins and outs of medevac and was set loose on the schedule. We had twenty-four- to forty-eight-hour ops on Oahu and, as I mentioned, a week-long op on the Big Island. I pulled a few duties on Oahu, and it was right out to the Big Island for the week-long duty. They sent me out with a low-hour PI, making me nervous as I wasn't entirely confident in everything we had to do. The duty requirements included flying the training area, conducting range sweeps to ensure no livestock was present, and then be on standby for medevac calls. Fortunately for me, nothing interesting happened during that week. We flew some training flights and our range sweeps, ate some good food that we had brought and made ourselves, and then returned to Oahu when the relieving crew arrived a week later.

While stationed at Hawaii I kept busy with maintenance, duty on Oahu, and whatever other training came down the pipeline. I thought having four MTPs for fifteen aircraft would make it easier, but the challenge of Hawaii was not having major facilities nearby to resupply parts as quickly as other units. We were the last to receive anything we needed. Additionally, with the humidity and saltwater, our aircraft were more prone to component rust, so regular corrosion inspections were more prevalent and time-consuming.

Also, with the limited number of full aircrews due to never being at full unit strength, we all pulled triple duty at times, or so it seemed. The joys and wonders of living in a vacation paradise were regularly tempered by the rigors of trying to keep up with the demands of flying and maintenance. We were lucky to have a CO who wasn't afraid to tell the customer "No" occasionally when maintenance wouldn't allow us to fly the requested mission. I never took a good CO for granted, as they were rare in the Army. The one who relieved him a year and a half later proved my point.

During our years in Hawaii, we traveled on family vacations to the Big Island, Maui, and Kauai. On the Big Island, we saw Volcano National Park and the Kilauea Caldera, traveled up to the visitor center at 10,000 feet on Mauna Kea, and drove to the northeasternmost point to overlook the lush, green valley below (it's the same location as one of the screen saver pictures on Apple TV), Hilo, Kona, some waterfalls, and sampled all sorts of local cuisine from a farmers' market. We also saw some turtles on the black sand beach on the island's south side laying their eggs in the sand. It was wonderful to stumble upon while making our way to the cabin we stayed in at the Kilauea Military Camp (KMC). We drove down to the ocean from there, but due to the hot temperatures and my girls' reluctance to do a ten-mile hike, we didn't venture to see the lava flow. Still, seeing the aftermath of what lava does to a terrain was an incredible experience.

We stayed a week in Maui at a resort on the Kaanapali Beachfront, a short drive from Lahaina, which has since burned down. The resort had five or six pools that fed into one another by a slide or a jump from one to the next. The girls preferred the pool to swimming on the beach. While there, we drove the Road to Hana, which features 620 curves, fifty-nine bridges, and numerous stops along its sixty-four-mile stretch, allowing us to see waterfalls and other historical sites or landmarks. We made it as far as Waianapanapa State Park to see the black sand beaches there, but with nightfall approaching and not wanting to drive that highway in the dark, we headed back to the resort. Jenny usually gets car sick driving switchbacks like that, as was the case this time. She almost made it through the entire highway, but we had to stop once so she could feel better. We also did a helicopter tour with Blue Hawaiian Helicopters while we were there. I have seen the north side of Molokai a few times, which features 2,000-foot cliffs with waterfalls that flow nearly year-round. I wanted to ensure that my ladies could see that. It's a breathtaking sight.

Since Jenny and I had been to Kauai, we were familiar with the island's layout. We only stayed again for a few days but had to take our daughters on the ATV tour that we had done nearly ten years prior. Again, it did not disappoint. When we went the first time, they had dune buggy-style

ATVs; this time, they had some high-speed side-by-sides that worked better for navigating through muddy terrain. We went over the President's Day weekend in February, which ended up being chilly, but we still had a great time. The midway stop to jump into the waterfall wasn't without its sharp inhale-inducing coldness, while not quite as exhilarating as a polar challenge. Even though my girls could swim well, they still wanted to hold my hand, jumping from the six-foot waterfall into the water below, and dragging me down while holding my neck as I tried to keep the air in my lungs. Of course, they wanted to go even more after the first jump, so I endured potential drowning for my daughters' enjoyment.

We also traveled all over Oahu. We frequented Wailua, Waimea, Three Tables, Sharks Cove, and our favorites: North Shore Beach and Sunset Beach, which had parking, picnic areas, and just the right mix of waves and water depth to play in. Banzai Pipeline was located just a bit further down, where the waves would reach twenty feet or higher in the winter. We would go there to watch the waves come in and go out, something we found mesmerizing.

We tried a few different Luaus. They were all great, with good food and entertainment. Everybody loves a fantastic Siva Afi show (the Samoan warrior dance of the fire knife), not to mention the other cultural dances of the Pacific Islands. We tried different foods, too, and soon developed our own list of favorite spots. Green World Farms was the best spot for coffee and tourist gifts. Kono's had fantastic food at a fair price and was in the more prominent city locations around the island. If there was a Kono's around, we were going! The Triple Crown was worth the cost; it would fill you up all day. Kalua pulled pork, ham, and bacon on a decent-sized sandwich roll. *All day, every day, twice on Sunday, please.* Pali Lookout was always a hit with friends and family that came to visit, and, of course, the *USS Arizona* Memorial.

Our favorite hike was Manoa Falls, which holds a dear spot in my heart because we visited it when we first arrived on the island. Charlize was six years old, and Izzy had just turned four. Charlize and I were a little bit ahead of Jenny and Izzy, when suddenly, it started raining. I asked Jenny if she wanted to turn back. She said, "We're here and wet; let's just keep going." Charlize was holding a small Minnie Mouse umbrella and doing her best to stay dry, unaware that we were already soaked. A bit further into the hike, she shook the water from her hand, turned to me, and said, "Daddy, my arm is getting wet."

"Chuck, the rest of us are soaked since we don't have an umbrella. Shake it off," I replied.

She looked at the rest of us, soaking wet, closed the umbrella, placed it in her hand, and kept walking with a smile. She had no complaint, just a smile as she was getting soaked too. We all chuckled over that one and went on to a gorgeous, heavily flowing waterfall! I remember being proud that, although I didn't have a son, a rough-and-tumble boy in the family, I did have two daughters who weren't so girly-girl that they couldn't stand being soaked, dirty, or enjoy the outdoors.

Finally, our favorite beach, hands down, was Kailua Beach. Gorgeous! The parking lot was located before a dune, so you couldn't see the beach from it, which we used to our advantage whenever we took people there. We would take our chairs, lunches, and everything since it was about an hour's drive from us. We'd stay there all day. When we arrived, whoever was with us would help carry everything, but we would let them walk ahead first. Once they got to the top of the grass dune, they usually stopped to take it all in. The warm ocean breeze would hit your face as everything came into view. Off-white sand and weather warn beach lines, leading to expanses of blues that started light and transitioned to a dark, navy blue, as far as the eye could see. Looking to the left was low-lying ground with the Marine Corps base and Kaneohe Bay, or K-bay, on the far side of the base. A high peak popped out at the end of the Marine base. There were a couple of small, cone-shaped islands to the right, to which people would canoe or paddleboard. The shore gradually sloped into the ocean, so the waves were always gently rolling in. The water was usually clear enough for snorkeling in the shallows to find items that fell from people swimming or paddle boarding. I found hats, sunglasses, and golf balls, and one time, a shiny object caught my eye; it was a small marijuana pipe. Lastly, there was plenty of shoreline, so we never felt crowded.

But it wasn't all about vacation and exploration fun. I was still drinking and at my worst, about a fifth a day on the weekends, half that during the week. I would buy a bottle and drink a few swigs between the half mile from the Class VI and my house. Drinking affected my sleep, and if I woke up early on the weekend, I would have a drink or two in the morning with my coffee before everyone else got up. I didn't do this every day, just most nights, as if that made it any better. We would pull a weeklong medevac shift on the Big Island, where drinking wasn't allowed. The longer duty also helped me slow down, but the Class VI was always on the way home and usually the first stop before getting home.

Still, I enjoyed the differences of the medevac mission. The purpose of medevac was to transport the wounded personnel, regardless of the circumstances, and provide them with the necessary care. We accomplished this by flying in and utilizing hoist operations, lowering a medic to help, and raising them or injured personnel to the helicopter. It was another type of operation to add to my skill set, along with FRIES, SPIES, rappel, cargo hook, and Bambi Bucket operations I had already trained in. The hoist had 250 feet of usable cable, but we typically conducted training operations to a maximum of about fifty feet. The highest I ever conducted a hoist op was seventy feet at night, which was unnerving as even a slight movement at that height could translate into a more significant movement some seventy feet below.

January 2017 was our turn for JRTC. Part of me was looking forward to it, but that anticipation was short-lived. I wanted to be a part of port ops, as I hadn't participated in that before, as well as the medevac operation in a simulated combat environment with all that entails. There was always a fuckin' LT to ruin every bit of it, though. We flew the birds down to Pearl Harbor to be loaded on a boat and shipped to Beaumont, Texas, where we would pick them up and fly them to JRTC—no issues, too easy. However, port ops automatically added ten days to two weeks to the process. No biggie, this meant hotels, or so I thought. We flew over to Beaumont, and our ship was stuck in the Gulf of Mexico for an extra five days, waiting for a boat at the dock that had a mechanical issue to be resolved. We didn't hate this idea because it meant we weren't stuck in JRTC for five more days.

Unfortunately, I had to fly with one of those guys who was a lost cause. He had about 400 hours in total, and while most people at this hour level were looking to make PC, he was still trying to fly straight and level in an aircraft with a flight director. I kept telling him to put the aircraft in trim and take his feet off the pedals; above 50 knots, the aircraft will maintain its own trim. He would leave his feet on the pedals and push them out of trim to make the entire aircraft fly straight forward, which would cause him to have to turn slightly in either direction to maintain a straight-forward flight.

On the first day we were flying into the "box," I noticed we were flying straight forward, but I was sitting higher than he was. Again, the aircraft was out of trim, and he was holding a slight right turn to maintain a straight flight path. I had had enough, and even while the OC-T was on comms in the back, I told him, "Put the aircraft in trim," which he did. "Take your feet off the pedals," I instructed him, and he pulled them back an inch or two.

"Put them on the floor!" I said.

"It feels weird to do that," he said.

"I don't care how it feels! Do it, and get used to it!"

He put his feet on the floor. "If you touch those pedals again before we get below 50 knots, I'm going to pull out my ASEK knife, unbuckle, jump across, and cut your fucking feet off mid-flight!"

He looked at me with a smirky smile as I stared dead at him, letting him know I was *not* joking. His smirk disappeared quickly, but he never flew out of trim again.

We asked the OC-T what he thought about that during the final AAR. He said, "No matter what goes on in the aircraft, I never break character or interfere so I can let it all play out. That day, though, I almost broke character and busted up laughing! I have never heard a pilot say that to another pilot, but I'm not surprised that it came from an MTP!"

JRTC was stupid, as it always is. We hadn't done much in the line of medevac operations, so our LT was trying to get some last-minute missions to validate our existence. I wouldn't have minded so much if he had told me, but as it was, my crew had the last duty, and this effin' LT decided to contact the BSO (battalion surgical officer, or something stupid like that) and have them request a bunch of medevacs. Starting at sunset, we had a medevac request every hour and a half. By 3 a.m., we were all pissed off about it. None of the flights was for a significant distance. We would fly to the medevac site, move them 500 meters to one kilometer away, and then return to the airfield. Shut down, nearly fall asleep, and get another call.

Was it good training? Sure. Flying multi-ship, low illum, a couple of confined areas, etc. Would it have been better if I had been given a heads-up about the whole thing? Undoubtedly. That's the difference between a good leader and this LT. His motives behind it were acceptable. A good leader would've let me know, especially on the last day, or gotten the three crews together and allowed them to share the load of his decision and get some flight time.

For whatever reason, the powers that be decided that all maintenance personnel needed to stay at the port until the boat left the harbor. Why? They don't need my crew or me to unload a boat. They don't need a MTP or crew chief for folded-up aircraft that they won't unfold until they return to Pearl Harbor. So, we were stuck in Beaumont for an extra four days.

I drank starting midday since we weren't restricted. I have allowed anger to be the cause of my drinking more than anything else, and since I was rather angry about the extra two weeks total that this trip had caused me, I looked for happiness wherever I could find it. I was glad to go home, but that would be short-lived.

"How was JRTC?" our CO asked me after I went back to work.

"It's over now. That's all that matters."

With a sheepish look, he said, "I'm sorry to inform you that your platoon will go to National Training Center (NTC), Fort Irwin, California, in August or September. I know you guys just got back, but you're the only platoon available, and this just dropped in our laps."

I went numb. Anger, disappointment, disbelief, rage; I'm not sure which one it was, or whether it was a combination of those and more. I laughed and said, "Great."

I shouldn't have been surprised that it was right around this time that I developed Bell's palsy for the first time. If you don't know what that is, your face looks like you had a stroke, but that's the only stroke-like symptom that you have. The nerves on one side of your face say "whatever" and stop telling the muscles to do anything. An entire side of your face droops. It's hard to eat or drink anything, and it's exhausting or painful to do so. There is no direct cause that anyone has been able to identify. The onset began the same day I was getting my DA photo updated for the promotion board starting in April. I have a copy of the photo, and you can see that my eye and lip on the right side have a slightly different look, with a bit of a sag kicking in, which serves as a reminder when I see it, although it isn't often.

I mentioned this was the first time; as the true overachiever that I am, I had it again in February 2021. When I first looked into it, I found that there was no clear reason why it happens, how long it lasts, or any other relevant data; it just seemed to happen. Stress appears to be the most significant common denominator of those who have it, and I had been under a lot of stress for the better part of seventeen years at this point. The chance of having it more than once in a lifetime is 20 percent. Check that block. Of those who get it, only 5 percent have permanent, lasting effects. Check. The first time lasted five weeks, cleared up, and went away, seemingly, for good. The only side effect was an occasional cramp in my neck, which was predominantly noticeable when I worked out or was straining at something.

After the second time, I had a noticeable difference in my facial features when I smiled or laughed harder than usual, neck cramps manifested on both sides when straining or drinking something rapidly, and I no longer had 100 percent movement on the left side of my face. I was tested for tumors, cancer, and anything else that could affect it; they all came back negative. It's just something I'm stuck with, and as long as I don't get full-on Bell's palsy again, I'll deal with it.

Another APART was completed in my aviation career, and I was headed to PTA in July, right after the Fourth of July weekend. It was a good deal since we only had to be out there for four or five days, but I was supposed to be going with the dipshit who I threatened to cut off his feet. Instead, I went with another guy who was a MTP and a W3, which made him the PC. I got to sit back and be the competent PI. It was laid back until the last night. Of course, it had to be the final night, and it had to be almost 11:00 p.m.

I was falling asleep—like when you are nearly asleep and starting to dream, though you aren't fully asleep yet. And then the house alarm went off. I slowly woke and thought, "Was that the medevac alarm?" I finished that thought when the medic came in yelling, "NINE LINE, NINE LINE, NINE LINE!" I jumped up and got dressed as the PC went to the phone. Once I was dressed, I told him I was headed to get the aircraft started.

I got the aircraft up to engine start as the PC approached the aircraft. He still needed to get his survival vest and helmet on before he hopped in, so I yelled, "I'm up to engine start with the rotor brake on; I'm going to crank them while you hop in!" He gave me a thumbs up, so holding the cyclic with my right hand, I reached up and started the first engine with my left. It fired up with no issues, and I cranked the other engine. "Starting two!" I said. Both engines were at IDLE. He was still getting strapped in, so I called "Clear the gear," and released the rotor brake. The blades slowly started turning, and once they reached the proper speed, I began advancing the PCLs to FLY. The PC was finally set and on comms, "A Marine took a ricocheted round to the ass and needs immediate evac. Weather is 3,000-foot ceilings and five miles of visibility the entire way down. Let's grab him, get down to the hospital, drop him off, get back, file our reports, and get some sleep," he stated.

"Sounds like a plan! Clear up left and right?" I bypassed his comments as I was ready to take off.

"Clear up left. Clear up right," the crew chief and medic replied.

"Coming up," I called out and we were on our way to save a life.

The Marines train with corpsmen, who are medics, so they had this LT stabilized before we even arrived to pick him up. The range was less than a mile from where we were, but the illum was low-to-nonexistent that night, and the terrain looked like tilled-up ground, from what I could see through NVGs. I landed directly where they had thrown some chem lights out. The LT was loaded up, and we were headed down to Hilo, flying about 1,000 feet, which lasted only a few miles. The weather report was wildly wrong.

First, it was low clouds. We could see lights in the distance, so I flew low and stayed under the cloud layer. As I continued to descend, I finally saw around 125 feet on the radar altimeter, which would place us about seventy-five feet above the power lines adjacent to the road. I didn't want to get any lower. Rain was also beginning to obstruct visibility. With clouds blocking any light from above, rain obscuring our vision, and a low altitude with obstacles present, I was focused ahead of us, spotting the lights for the hospital's area, when a gut-sinking feeling hit. We were five miles from the hospital, with a wall of clouds directly in front of us. I couldn't go under; they went to the ground. I turned left to go north to skirt around the clouds for an opening, but nothing. The whole time, the PC was on with

ATC to get them to vector us around and into the hospital. ATC was no help whatsoever.

I began a climbing left turn to go over the clouds using the flight director to keep everything on track. This is when I realized how lucky I was that I was not here with the incompetent pilot I was supposed to be with. ATC finally placed us on the approach plan for Hilo airport, which would get us down to 400 feet by the end of the approach. The only downside was that the approach took us fifteen miles out over the ocean before it brought us back in. We pulled the guts out of the aircraft at this point as there was only a patchy cloud layer over the ocean.

After making the right turn to crosswind, I could finally spot the runway. I knew the hospital was just beyond it and to the right. I spotted a cluster of lights and announced that's where I was headed as the PC told ATC that we had the hospital in sight and were heading directly. We circled the hospital, and I spotted the helicopter landing pad. Since we were close to it but high, I had to do a steep descent, a swooping left-hand turn and land at a high speed similar to a dust landing. Luckily, it was barely above sea level. We were lower on fuel and weight and had a significant power margin to pull for the landing. The PC and I fist-bumped on a job well done as the crew chief, medic, and corpsman took the LT to the ER.

We returned to base with just enough fuel in reserve, talking about the whole ordeal all the while. We were on the phone with our flight folks and medics until nearly 4:00 a.m. when we told them we'd deal with it all when we got back in the afternoon since we all needed some sleep. They agreed and let us do it. The relieving crew arrived, and we shared the story like some guys returning from combat. They couldn't believe the scenario of someone getting shot, the weather, ATC, and everything we had to endure to get the LT to help, but they were glad it all worked out. This wasn't the first time in my military career that I felt like I did something valuable or worthwhile, but it was the first time in my aviation career that I felt I did something genuinely purposeful. It felt amazing to know that I was a part of something that helped to save a life, even if only one person's life. Even writing about it now, years later, I am reliving some of the emotions, and in the end, it still feels great.

Emotions driven by circumstance can run rampant in the military. One moment, you are on a high; the next day, week, or month, it all comes crashing down. Aviation mishaps are infrequent but when they do happen, they shake the entire military community, especially on a smaller scale, like in Hawaii.

I had met Steve Cantrell at AWOAC a year before he came to Hawaii. He was an intelligent guy, somewhat funny, but he irritated the class as the guy who always asked questions when the instructors said, "Are there any questions?" Of course, this would result in a five-to-ten-minute answer to his question when all we wanted to do was leave, and most times, he would have another question that would drag it out even further. He came at the same time as another guy I met and befriended, both of whom were MTPs, both of whom came to medevac. Both coming to the Med gave us too many MTPs; one would need to go to the Assault unit across the airfield. Steve was the one who ended up going to the assault unit. Steve had conducted most of his progression flights with the Med before he went to the Assault. He only needed to complete the nighttime iterations of whatever he had left to do.

A common practice among SP/IPs is to take whoever needs training flights for the day or night, throw them in the back of the aircraft and swap out trainees as required.

Loading all trainees on the aircraft reduced the need to return to an airfield and pick up other pilots or crew chiefs. Crew swaps could be performed as needed for training purposes. Steve was in the back in one of the passenger seats; the SP and PI were flying, and two crew chiefs were at their stations. I heard that the PI was a low-hour LT who wasn't comfortable flying formation flights with NVGs, especially under low illum conditions. Another factor is that flying over the ocean doesn't give you a solid view of any horizon, so if you aren't looking at your instruments, it can be challenging to tell if you are flying straight and level instead of at a slight bank or diving.

Either way, the PI was on the controls while the SP was making a radio call, and they were flying chalk two in formation, over the water, low illum. These are all the ideal criteria to help the LT work through her fears of flying in this environment. While the SP was focused on the inside of the aircraft, making a call, the LT announced, "You have the controls." The SP immediately took the controls and somehow placed the aircraft in a slight dive. He didn't recognize the aircraft's attitude and further nosed the aircraft over. When he took control, they were at 1,200 feet and began a rapid descent. They were unable to correct the aircraft's attitude or regain control, and it struck the water at approximately 180 knots with a 30-degree nose-down attitude. In the recording, you can hear Steve say, "Pull up, pull up!" It was too late.

I woke up the next morning to numerous texts and a couple of voicemails.

"I just heard what happened out there, wanted to make sure you're OK," one voicemail stated from someone I hadn't seen in three years.

"Do you know who it was?" a text asked.

I had no idea what had happened but from the messages I knew there'd been a mishap. I received another call around 5:30 a.m. that I was awake for.

"Hey Chris, we may need some crews on standby for search and rescue. One of the assault aircraft went down over the water last night. That's all the details we can share at the moment."

"Sounds good, I'll be to the hangar as soon as I can," I replied.

I got dressed to head into the hangar in case they needed me to fly but whoever was on shift would take the first flights, I would be backup to the backup. I didn't rush. As I got dressed, I pondered what had happened, who it was, and whether they were still alive out there, floating around, waiting for rescue. I highly doubted it. I knew this could happen to any of us at any time but still, when you hear about it, you're grateful it wasn't you but also sorrowful for the family that has to carry on.

"Hey babe," I said, gently waking my wife. "I'm headed in to work now."

"So early?" she asked, stirring some but not waking completely.

"Yeah, a crew went down last night; they want some crews around for standby."

"What happened?" she inquired, more awake now with the same thoughts probably coursing through her mind.

"I don't have the details, but I'll let you know more when I find out." No real reassurance from my end but I didn't know.

"Oh . . . OK . . . well if you have to fly be safe, please!" she said as she rolled over, trying to go back to sleep. I doubt she was able to.

"Of course," I said as I headed out the door to find out what I already knew.

As I've often heard in my Army career, "The train never stops." A week later, we were in San Diego conducting port ops for my first and last visit to NTC. Port ops went quickly, and they sent us up to Fort Irwin immediately after.

I preferred the NTC setup over JRTC. We weren't required to wear body armor unless we were flying, which was a nice change from JRTC. We were all fresh off the recent fatal accident. During training one of the aircraft was damaged here, too. The Assault company was conducting dust land training with some of the PIs, and a harder-than-normal landing resulted in the main rotor bowing downward so much that it chopped the top of the tail boom and severed the tail rotor driveshaFort All this while the unit was holding a memorial for the fallen a couple of weeks prior. While the train doesn't stop, more than a few of us mentioned that it should hit a terminal and take a breather. It never did, but fortunately, no other incidents occurred here during training or in day-to-day operations back home.

The fix, or risk mitigating factor they came up with after the accident, was to add a block to the risk mitigation matrix on the RCOP for dunker qualification and currency. Dunker is the name of the course for the military's underwater survival training for aviation personnel. It's required to have the training completed every five years if you are stationed in a unit that routinely flies overwater. While it is good training, none of it would have helped the crew that hit the water at 180 knots, with the nose of the aircraft about 30 degrees down as well. There are Band-Aids on bullet wounds, and then there is whatever the hell the decision to add this to the RCOP was. Dunker wasn't saving anyone that day.

Of course, NTC or JRTC wouldn't be the fantastic training it is without a nice round of headbutting with the OC-Ts mixed in for good measure. My crew was up for a medevac, which should have been relatively easy. Someone at our site was "injured," I say that with quotes as it was notional, and they simply needed to be transported to a higher level of care. We were to transport them there. No rush, but we still treated it like the real deal happening in real life. We went to the aircraft, loaded the patient, fired it up, got the blades spinning, and ensured we had the most up-to-minute maps before takeoff. The restricted operating zones (ROZs) changed regularly around NTC, as units could conduct live-fire operations from their locations due to the vast size of the training area. The OC-T in the back had an app on his phone that allowed him to view the ROZ information and track our movements while in the aircraft He used this not so much as to call us out but to keep us, and himself, safe from potential dangers.

However, his phone was delayed in updating his location that day. As I took off, he said, "You need to come to the right some more."

"That will take us into a ROZ," I said.

He replied, "You're headed to the middle of one right now!"

I couldn't see his phone, but he couldn't see my screen. I was splitting the difference between two ROZs, and he was trying to turn me into one.

I began the right turn, and shortly after that, his phone updated, and he yelled through the internal comms, "TURN LEFT! TURN LEFT!"

I thought I was about to hit something in the air—my heart rate jumped, I was having a few cold sweats through the panic—and turned a sharp 90-degree turn to the left.

Apparently, that was too far as, within a few seconds; he yelled, "TURN RIGHT!" Now I wondered what the real threat was: small red circles on a map, another aircraft, or this asshole E-7 and his phone! Finally, I brought the aircraft back on my original course and told him to look up at my screen. He saw the flight path, and reluctantly agreed that it was a safe course to fly.

I blasted to the entire crew, "Until you see another aircraft flying directly toward us, and it looks like it will hit us unless one of us alters course, everyone needs to SHUT THE FUCK UP!"

A moment later, one of the crew chiefs calmly and quietly said, "Aircraft, same altitude, nine o'clock, headed in our direction."

"Thanks, man," I said. We continued to our destination, dropped off the patient, and returned.

"I'm sorry for the mix-up. It won't happen again." I don't remember what I followed that up with, but of course, I was pissed off and took it out on the OC-T.

There wasn't much more to talk about NTC. We headed back to San Diego and loaded up our aircraft, but instead of putting us on a plane and sending us home, they put us on a bus back to NTC. It was cheaper to get a larger plane to take us back to Hawaii from the airport near Fort Irwin than to send us home commercial from San Diego, or so we were told. I don't believe it for a second. We had three more days to hang around and do nothing until we went home. I don't see how this was more cost-effective. Meanwhile, anger rose, and the desire to drink increased.

We had to get up at 3 a.m. to catch the plane, so here we were, a plane's worth of soldiers, lined up with all our bags, waiting for buses to take us to the plane. Then the buses were late. Allegedly, the bus drivers were never informed of the showtime, so they were five hours late picking us up. No matter; the aircrew wasn't there by the time we finally got to the tarmac, because they had another hour or two for crew rest. We had five buses full of pissed-off people who could've been home three days prior if not for "cost-effectiveness." The flight crew finally arrived, but the plane had a mechanical issue. Luckily, they had the part at the airfield, adding only another hour to our wait. It was nearly 2 p.m., and the only saving grace was that we would be traveling back three hours and land at the same time we took off in Hawaii time. Everything combined, I hit the Class VI on the way home, determined to maintain control of my drinking, which I did for a time. And then, Pacific Pathways.

CHAPTER NINETEEN

Pacific. Fuckin. Pathways. Or PPW for short. It's supposed to be a political handshake, or so we were told. A way for us to train with the forces of other countries throughout the Pacific to further strengthen relations with these countries.

In early December 2017, PPW fell apart. We all hoped we wouldn't have to go, especially after our recent trips to JRTC and NTC. Funding was approved after the new year. The aircraft were loaded the first week of January 2018. We had a small Task Force command of four personnel, led by a major, and we were set to leave on February 5 for a four-month deployment. I was committed to sobriety, and this would test me further than my commitment.

On February 5, 2018, we boarded a flight headed to Thailand.

After a quick stop in Japan, we arrived. I thought Hawaii was warm in February, but Thailand was a punch in the face. I can only imagine what it would feel like mid-summer. Port ops were even worse, as they were on the water and had the added joy of humidity to everything we did. This is where it went downhill rapidly. Every other time a unit had gone to Thailand for PPW, they stayed at the Ambassador Hotel, but not for us. Nope, they crammed us all into some tiny, open-air, piece of crap building with no comfort items except what we brought and a cot. It was ridiculous to see ceiling

250 • CHRIS KRUGER

fans that resembled old oscillating tabletop fans, attached
to the ceiling every fifth tile. Only half of them worked, and
they didn't help to circulate the air anyway.

I was already nearing the peak of what I would learn
later was traumatic stress, coupled with off-again, on-again
alcoholism. My way of dealing with anxiety, if not with
alcohol, was anger and angry outbursts. I wasn't sleeping
because the whole room was a snore-fest, combined with
the constant pinging of people's phones from receiving
WhatsApp messages. I wanted to kill everyone who didn't
place their phone on vibrate. Seriously, after every ping, I
imagined walking over with a hammer or whatever heavy
object I could find, smashing their phone, and, hopefully,
they would lunge at me, and I could bludgeon them, too. I
needed help.

You may be thinking, "Chris, that doesn't sound so bad.
You're in the military. What do you want, the Four Seasons?"
Yes and no. All I ever wanted was to believe in the delusion
that someone in the chain of command beyond the company commander cared about us. Mission first? Sure! No
doubt! But to what extent?

Another factor at this point was during my eighteen years
I had already deployed to the Middle East for longer than 70
percent of the people around me have. What's worse? I'd already deployed twice as many times and had twice as many
lengths of deployed time as those who have been in the
same amount of time and were also in charge of me! Can you
imagine how infuriating it is to have eighteen years of military experience, eight years of aviation experience, and four
years of MTP experience, and then have all your suggestions,
recommendations, and experience straight up dismissed by
some fuckin' LT that has been in for three years? It's beyond
infuriating. To take a training or live scenario, lay out every
possible course of action, recommend the best one to take,

and then to have every bit of advice–based on years of experience–is disregarded because some fucker has a college degree in criminal justice is just criminal in and of itself. If some of this sounds like entitled bitchiness, then I'll concede that some of it is, but I think I have well and truly earned the right to it at this point.

Digressed tirade complete. Let's continue.

I will admit that flying around Thailand had some beauty to behold, what we saw of it through the regular smog in the area. Picturesque, jagged tan rock formations sporadically covered in green trees or shrubbery. A few of the pictures I took of the landscape while flying around looked like screensavers preloaded on a computer, though, through the wonderous landscape there was always the presence and smell of smoke–wherever that came from. The living conditions in the staging area for our training weren't any better. The only difference was that we were spread out and had our company area.

Aside from routine maintenance, we sat and waited for a medevac call. We didn't do much flying, and neither did the Assault. They were more of a taxi service for the "bigwigs" that bothered to come in for some handshaking appearances. A couple of senators and congress members came. I can't be too sure, as I didn't give a flying fornication about them or what they were doing there. I didn't care for Thailand much and was happy to leave. Maybe it would have been a better place to visit if not for the Army BS we had to endure, but I doubt I'll ever go back.

Port ops on the backside ended up being a kick in the balls too. Our PSG volunteered to head out last as sometimes staying longer came with perks, like a hotel stay before heading to the following location. However, this time, with these clowns, it backfired on us. We had an extra five days to hang around in that same shit building, eating the same shit food, with nothing to do since our birds were already on a boat on the way to South Korea. We were approved to stay in a hotel if we wanted, but we had to arrange our transportation and pay for it ourselves. Aside from our PL and PSG, the six remaining decided it was worth the cost and got a ride to the Ambassador Hotel.

I just wanted AC, a pool, and good food. The PPW mission was titled Cobra Gold and I was irate when I walked by the hotel's restaurant and saw signs pointing to the dining area that stated, "Welcome Cobra Gold." The hotel was so used to us staying there that they had signs directing us to the restaurant and other facilities—just another kick to add to the balls. Who cares about soldier morale? Mission first, remember, not morale. But I'm here to tell you that the latter can drive the former to tremendous success.

We hopped on the bus to the airport to begin our journey to South Korea. Of course, there was a delay in the bus taking off. It was 100 degrees, the AC on the bus wasn't working, and for the next forty-five minutes, we were all sweating like stuck pigs, which we essentially were. We finally arrived at the airport, where we were promptly charged an outrageous fee for all our baggage and were on our way to South Korea.

We landed in South Korea at 22-degree temps, and I was all for it! I hadn't had cold temps for over two years, and I was ready for the need to wear cold weather gear again. We hopped on a bus for Camp Humphreys, our first holding area, until our aircraft reached port. We stayed in some unused barracks for the first few days, a welcome change. We were right by an older mini-mall with some food offerings, a movie theater that hadn't closed yet, and a few other amenities. We tried to enjoy what we could, knowing what lay ahead.

South Korea already has Army Aviation units throughout the country. We had to undergo Local Area Orientation

(LAO) flights and be signed off, ensuring we were familiar with all the areas we were allowed to fly into, as there were plenty of places we weren't permitted to fly into, the DMZ being a major one. All these things combined meant that they didn't need us there, and worse, they had nowhere to place us. We knew we would be bouncing around from place to place for the next two months until we headed to the Philippines, and it was beyond frustrating to have to deal with it.

Somewhere amid the sleeping on cots in open bays next to annoying people who didn't take personal calls away from the group and lack of sleep due to phones going off all hours of the night, I had a run-in with a senior E-6 who thought he was more important than he was because he was the quality control NCO in charge (QCNCOIC). I received a call that one of our crew chiefs had dropped a screw bit, and it landed somewhere around one of the tail components. He was an E-4, and SPCs are known for having attitudes that far outweigh their rank. This SSG happened to be there and asked the SPC, "Are you going to get down and find it?" To which the sarcastic SPC replied, "Nah, I figured I'd just leave it where it fell." The SSG corrected the situation, and that should've been that. Instead, he wanted to make a big deal of the whole thing and informed the PCOIC and PCNCOIC (production control officer and NCO in charge, the head of maintenance). I received a call from all three people, but no one told me who the soldier was. Knowing my guys, I had two in mind who would reply to an NCO in such a way.

I informed them all that I would handle it, but before I did, I wanted to meet with the QCNCOIC SSG to find out who it was, so I could address the issue internally. He agreed to meet me in thirty minutes outside the gym where we were staying. Thirty minutes later, there was no SSG. I texted and called him, no answer. Now, all the anger I had toward whichever SPC did this egregious act (sarcasm, I didn't care that much but was going to put whichever one it was in their place) was redirected toward this SSG. He was wasting my time within something that I already considered a massive waste of my time, meaning PPW in general. I was not having it, none of it! Forty-five minutes later, nothing. Finally, fifty-three minutes later, he texted and said he'd meet me outside, across the street, at the subway. He was getting dinner. *I'm about to wreck his dinner.*

I walked up, and he started to say, "Hey Sir—" but I cut him off mid-sentence and said, "Who was it? Give me a name." You see, I was ready to go tear up whoever it was, but then the SSG fucked up.

He said, "I don't know who it was, but it was a Delta Company crew chief."

Delta companies are maintenance only. They don't fly; they do ground maintenance, so it wasn't even one of my guys. I blasted into him, "So you wasted *my* time to tell *me* about an incident that didn't even pertain to one of *my* soldiers?! You tell *me* that *you* will meet *me* in thirty minutes, and now, nearly an hour later, *you* finally show up to tell *me* that it wasn't even one of *my* guys!" He tried to interject, but I cut him off with louder yelling and a knife hand gesture at him. "*You* must have recently been promoted to lieutenant colonel! Maybe I should be standing at attention for *you* and addressing *you* as Sir!" His face was red out of embarrassment, and mine was red out of anger. I was shaking out of pure anger and what could only be described as hatred for this guy as I continued to rip into him. I don't remember what I said for the rest of the time. Still, I finished it more calmly than I started by saying, "If *you* have any more issues with *my* guys while we are here, *you* come to *me* directly. Do not talk to any of them anymore, understood?!"

"Understood," he fearfully replied as I turned to walk away.

I needed help, and I knew it. This wasn't the first time I had an angry outburst far above and beyond what the situation called for. The assault unit guys heard about the incident and how I tore into him, which was salt in the wound for me. They all thought it was hilarious because they didn't like him and thought he was overstepping his bounds as QCNCOIC. It made me feel a tad better even though I instantly felt horrible about how I handled it and let my emotions run wild and drive my behavior. I wanted to apologize to him but couldn't bring myself to do it. No worries, God provided that opportunity a few years later, and I took it.

Emotionally, I was all over the place—all the previous circumstances and then a call from my wife. Our rental property in Fayetteville had been broken into, and the tenant was assaulted, or so the police report stated. All this while the HVAC unit was on the fritz, so we tried to get all that taken care of while our property manager was in Fayetteville, my wife was in Hawaii, and I was in South Korea. There was some confusion in the email traffic, and the repairs took longer than necessary to complete. Add to that a sliding glass door that needed to be replaced. We suspected the claim's legitimacy, but the house wasn't in the greatest of neighborhoods.

I was snapping at anyone for the littlest thing and would find myself outside, in the cold, pacing when I couldn't sleep because of someone snoring, their WhatsApp pinging, a late-night conversation, or any combo thereof. I wanted to light the world on fire and be found holding the matches. I wanted to die, sometimes, or go to sleep and not wake up. We had life insurance, and my girls would all be OK. I just wanted never to have existed, period. Some of me wanted to project that onto others, and I was even more miserable. I'm not sure it would have ended well if it hadn't been for a few close friends on that deployment. We kept each other sane when the weight of the situation was getting overwhelming.

We wrapped up our time in South Korea with only one small training event, returned to the port, and then headed to the Philippines. Another port ops, another bit of flying to our destination for the duration of PPW. It would be a total of three weeks, and we eagerly counted the days.

After returning to Hawaii, I had one more year left before our next PCS, and I wouldn't be on another training or PPW deployment. I determined that if I somehow ended up on something like that and had to leave for longer than a week, I would fight it with everything I had. Some guys get stationed in Hawaii, don't leave the island for three years, and then move on to the next duty station. I wasn't that fortunate. For now, the Philippines.

We set up in a dump similar to where we had stayed in Thailand. It had an open floor area—no ceiling fans this time—and we slept on cots enduring even worse food and bathroom amenities than we'd had in Thailand. We were told that AC units were on the way, but then those got scrapped due to a budgeting issue. Then, it was $10,000 worth of industrial fans coming to help circulate the air and make it a bit more comfortable for sleeping, which turned into one large industrial fan for each area we had; there were five. Getting your hopes up for something we were told was coming is one thing. It's another thing not to know and have it show up. It was four months of having the proverbial rug pulled out from under us, and my head was starting to hurt from all the hits. Plus, all the personal shit going on. Then, the icing on the cake.

I was scrolling through social media one day, and I saw a buddy post two drinks and one of our Fox Company (F Co) LRS hats with the caption, "One for the Airborne Ranger in the sky. Rest in peace, Michael Blizard." It was like a punch in the gut. I didn't know, but he had gone into Hospice care; I assumed it was because of a lifetime of drinking and smoking. He had just retired within the last few years, and I'm glad I had gone to see him once when we were visiting family in Fayetteville. He was the 1SG at the Air Assault School they opened at Fort Bragg, and we caught up for a good hour or so. He was smoking then, too, of course, and I'm sure if he wasn't at work, he'd have had a beer or whiskey in his hand as well. He wasn't so much a mentor, even though he was our PSG and 1SG during my last years at F Co, but he was a friend. I say he wasn't a mentor because I didn't learn much from him; he was a friend, as we saw eye-to-eye about the unit, the mission, and the deployment. His son, Chase, committed suicide a few years after his father died.

Devastated over the news, we finally received some that brightened my day. They needed two medevac crews to head north to support some training. We would head up there with two crews of the assault unit to be ready to launch medevac in support of the operations they were conducting with Marines on one of the islands in the northwest portion of the Philippine islands. It also meant we would be set up in some bed-and-breakfast hotel—AC, showers, three people to a room, and three cooked meals per day.

I was relieved we could be somewhere with some comfort for the rest of our time. This meant we only had to be in a shit box of a place for two more days before port ops, a hotel, and a plane ride home. If we weren't watching movies, I was sleeping and trying to get rest from our troubles while I could. We weren't allowed to conduct any training flights, except NVG reset flights, and I didn't care in the slightest. I was ready to go home and stop wasting life on this waste of life!

Fast forward to port ops being complete, and we headed home. Just like NTC, the plane was late, and we were all hanging out for the day in a hangar on some random airfield in the Philippines, waiting for our ride home. I did my best not to let it bother me, but I was starting to fail at that. A plane finally arrived, and we were able to get home. We were then granted a four-day weekend before heading back to work.

We returned from PPW in the first week of June 2018. The first thing I did after getting home, greeting my ladies, eating some good homecooked food, and spending time with my family, was run to a bottle. I was infuriated by the waste of the last four months. I was unhappy about the time that had gone by, which I would never get back. I knew I was also moving to a new position, and that stress hit me as I wanted one position over another. So, I drank. My sobriety was gone again, and I'm sure I spent most of the weekend in a drunk, or at least buzzed, state. When I went back to work, I called Behavior Health to say that I wanted to start treatment, and while talking to the CPT I would be seeing, we set up an appointment. I never asked if she could tell, but I was holding back or wiping away tears the entire time we talked.

If the first step is identifying that you have a problem, then I knocked that one out of the park. If the second step is getting help, that was harder than the first because it felt like admitting I couldn't handle it alone. I thought that having faith alone would be enough to get me through it. Maybe if I prayed hard enough about it, had enough self-determination, and did enough of the right things, I could make it through on willpower alone. I was wrong, and how! Even great men of faith need help to persevere. We weren't created to be individual and self-sufficient. We were created for relationships, which was what it would take to get through.

For the first time in my aviation career, I decided to take control of my path. I was supposed to take over the QCOIC position, which would be a cushy job, allowing me to dictate my schedule again. With no one giving me any real insight as to what my new position would be, I packed up my stuff, moved into the QC office, set up shop, and got to work. No one questioned it, not even the BN CO, who should have been the one to tell me what I was doing next anyway. I attended the required meetings and continued introducing myself as the QCOIC. Everyone went with it. It was the only time it worked in my favor, though. This was my way of trying to regain a little sanity.

In mid-June, my mental health journey began. I was going to counseling weekly, and, in the meantime, I was placed on a temporary down slip so that I could focus on counseling without the distraction of flying. The down slip was an addition to the counseling, not a precaution, and I was grateful for it. I wouldn't be bombarded by a random MTF that needed to be done ASAP or to troubleshoot an issue on some random broken aircraft

I also sought a sleep study for sleep apnea, as I had a few sleep issues: snoring, snoring that wakes me up, gasping for air numerous times in the night, at least two-to-three times a night, and waking to pee multiple times. I had constant allergy-like symptoms. I had aches and pains all over, all the time, and Tylenol or ibuprofen wasn't cutting it anymore. Lastly, the flight doc wanted to remove a couple of moles I had, which looked malignant—they weren't. As I was getting all these medical issues checked, I also thought it a good idea to take some leave. Since I had a decent amount saved up, I took a whole month off to relax and get away from it all.

Mid-July, counseling was going well. I was not drinking too heavily, but I knew it needed to stop. I wanted a change, a life makeover change, and I knew I probably never would if I didn't commit now. My birthday was coming up, but I didn't want to wait until after, as most people do. July 18, I decided to reengage in my sobriety. I finished off whatever booze I had and had a good buzz going, but I felt empty, as had often happened.

I know the cycle of alcohol, feeling a buzz, happiness, euphoria, and that's what I would try to maintain. Then, drink too much, realize I'm getting drunk, not give a damn after that, have one more, and wake up the next day feeling like a trainwreck, looking like one too. I wanted off, and I wanted off for good. For my birthday, we went on an ATV tour at Kualoa Ranch, and while it was fun, I had a hard time being hungover, though I hid it well and my family was none the wiser. I used that as a reminder for the near future of something I didn't want to walk back into.

On July 20, I committed to working out every morning for at least thirty minutes. I was nearing 190 and wasn't having any of it since that was the heaviest I had ever been. We had a detached one-car garage, and I had some kettlebells and adjustable height hanging rings. I was determined to get back to 180 pounds through bodyweight exercises, supplementing with kettlebells when I needed more weight than my body could provide. I would work out four days weekly and, on Wednesday, have a run with an abdominal-focused workout. By sticking with this, I was back down to 180 pounds in about a month. We had a PT test in early September, and I was pleased that I scored 296 out of 300 within a year of turning forty.

I completed my sleep study in August as well, and it determined that I had sleep apnea. I would be receiving a CPAP machine soon, which I would discover is a pain in the ass, and I hated using it. It didn't seem to help me much, but I used it anyway. In December 2018, I was back up and flying again. The QCOIC position had been a great hiatus; I was becoming holistically healthier, and our marriage was doing well. Life was good. And then, Human Resources Command (HRC).

HRC always came to Hawaii in person, usually in December or January. This time, it was the first week of January. I listened to all their usual BS, including promotion rates, aviation bonuses, how to play the aviation game, and so on. I was only interested in where I was going next, and this was the time to find out. I was at a crossroads in my military career; this time, it pertained to my retirement.

I saw three options. One, I retire at twenty years. This option meant the next duty station needed to be somewhere without a commitment, and Hawaii had a thirty-six-month commitment. Two, after getting to the next duty station, I take the aviation bonus that I would be eligible for; we ride out the three-year ADSO (active-duty service obligation, a fancy way of saying, "I gave you something now you must provide the Army more time"), I retire at twenty-two-and-a-half years after that, and we move on. Three, we push that last scenario to twenty-five-and-a-half years as I would be eligible for W4, which would incur another two-year ADSO, and we'd have a bit more banked for retirement.

I sat down with the HRC branch manager, who had been stationed in Hawaii, and the GSAB with us before taking this position. He had the great honor, nay, the privilege, of informing me of my options for the next duty station: either South Korea or Fort Drum, New York. He also stated that it was between another W3 MTP and me. We would each be going to one or the other of these places.

I already knew that Drum would be deployed in late 2019, and while Korea would be a neat cultural experience for my family, I had little desire to drag them all, including a cat, to South Korea. I informed him that I would take Drum, and he said some crap about having to weigh options between me and the other guy. I said, "Well, I'm here talking to you now; doesn't that give me a little leverage?" He agreed that it did, which it must have since I had orders about a month later for Drum.

That last year in Hawaii was the most peaceful one we had while I was in the Army. I went home early every day; we enjoyed the island as much as possible since we didn't know when we would return. We just spent as much time together as a family as possible, knowing that we would get to Fort Drum, and within three months I'd be off to Afghanistan again. I wasn't as downtrodden about the looming future as one might suspect. I knew that no matter what, it would be my last deployment.

We headed for Baltimore, Maryland, on June 10, 2019, three years on the dot. We were exceedingly fortunate that Jenny's brother was stationed at Fort Meade and had room in his house for us. We figured it would only be a couple of weeks until Jenny's car arrived at the port there, and we could have our house on post, with our household goods coming as well, and move on with our lives there. We stayed with him for five weeks, and in July, we headed up to Drum and our new location for the next three years.

Fort Drum. You either love or hate it; there isn't an in-between. Growing up in Spokane, I was used to snow in the winter, and I missed it. Our girls were excited about the possibility, and my wife was on the fence. Before we deployed, I bought a snowblower and snow tires for one of our rides, and the girls all had a good time picking out their first snow gear. They wanted to try everything while we were there: skiing, snow tubing, snowmobiling, everything you can do. But first things first, I had this deployment to get out of the way.

In October 2019, I arrived in Kandahar with the torch team—eight total. We went earlier than the others to arrange their arrival and began the transition process. After I had preflighted my first aircraft for our LAO flight in our flying area for the next nine months, I could tell the aircraft looked horrendous! The blades were bare with no paint, the top layers were peeling back, and the spar, the leading edge, looked like it could separate from the rest of the blade at any time. I looked down the line at the rest of the aircraft from where I was atop one aircraft, and they all looked the same. Shaking my head in disbelief, I knew our work was cut out for us.

I signed into A Co 2-10, Air Assault, Voodoo. I was relieved to be back with Air Assault after my time with Med, I wouldn't be the only company MTP, and the other MTP was a W3 and even more capable than I was. Before becoming a pilot, he was a Chinook crew chief, so he was well-versed in all the bookkeeping stuff I mainly left to the crew chiefs. I could grind my backside off and get things done rapidly and efficiently, so we made a great team. We posted a sign outside our office in Kandahar, Afghanistan of Riggs and Murtaugh from *Lethal Weapon* movie fame. That's how close we worked, and it was fitting. That was one of the things that made it a great deployment.

Shortly after the redeploying unit departed, and all our personnel were present, we discovered that all the engines were in poor condition too. These aircraft had been there so long that we were limping most of them along through the deployment the best we could. No matter. Again, I wasn't alone this time and was beyond grateful to have another MTP with whom to share the load.

Most of our maintainers were inexperienced, and that began to show quickly. Maintenance tasks took twice as long to complete, and mistakes were often made due to inexperience or incorrect reading of the steps in the maintenance manuals. Items were improperly reconnected or not connected at all. Anything with liquid in it was over- or underserviced routinely. Finding someone who could assist me with anything I needed in a time-efficient manner was a challenge. Flushing an engine is a fundamental task, but if the crew chief didn't know how to hook up the wash wands or prep the flush cart, it took extraordinarily longer than it should have. I found this drew the anger out of me that hadn't been present in over eighteen months since PPWs.

One day early in the deployment, an aircraft needed an engine flush. I instructed the PSGs to gather everyone who wasn't busy and have them come out to the helicopter to observe how it's done, in order to facilitate a more time-effective solution in the future. They agreed, and we had about a dozen guys ready to learn. One of the E-5s walked them through how to hook everything up and prepare the cart, and we were ready to flush. I didn't see that he had two sets of hoses, one hooked up to each engine. Usually, we only had one set of hoses, and the crew chief had to bounce back and forth between the engines, disconnecting and reconnecting the quick-release fittings between flushing and rinsing. Either way, my not noticing that was the least of the problems. Before I started an engine, I got everyone around the flush cart to show them which position the two sets of handles needed to be in for operation. Switch it to the chemical setting for the flush and water for the rinse. Open to let chemicals or water through, and bypass to stop the flow. Simple, right? I set it to chemical and bypass and informed the first guy that I would tell him on the radio when to switch from bypass to open and give him a thumbs up so he had both a verbal and visual signal. I started the engine, got it up to idle speed, and informed him to start the flow. He turned the handle from chemical to water and then looked at me. Mind you, he was about thirty feet off the nose of the aircraft, so I said, through the mic that went into the headset he was wearing, "What are you doing?"

"Turning on the flow," he said.

"No, you turned it from chemical to water, and nothing flows." I also had only two minutes to accomplish this task before I had to turn off the starter so it could cool down. He reached down and moved the same handle back to chemical, after which I said, "Now turn on the flow!" He immediately moved the same handle back to water. I said more forcefully this time, "Turn it back to chemical and turn the *other* handle from bypass to *open*!" He moved the handle back to chemical, then stared at me. I realized at this point they put the dumbest motherfucker on first, and I lost my shit! "NOW TURN THE OTHER ONE TO OPEN!" You can guess what he did. Yup, he turned the same handle back to water.

I have no doubt that I came off as an asshole as this whole thing progressed over ninety seconds or so, and I was shutting down the engine, and the pilot that was in the aircraft with me learning how to do maintenance tasks was dumbfounded too. She couldn't believe what had just happened. As I shut the engine down, I got out, grabbed up everyone who stayed for the shit show, brought them around the flush cart, and went through that whole block of instruction again.

"Do you all understand now?" I asked and got silent nods in agreement. Secretly, I wondered if they did.

Frustrated, I sent this idiot away along with the rest of the people and turned to grab the E-5 to finish it up. He was storming off to another aircraft that needed a stabilator repair due to some dumbass crashing into it with one of the gators that we use to haul equipment up and down the flight line. He sent me another guy to do the flush.

"Where is the sergeant?" I blasted, growing even more irate than I already was.

He returned around the corner and said, "I don't like how you handled that, and I'm not going to finish the flush or work with you."

I lost my shit. I stood nose to nose with him, leaning in like a drill sergeant would, causing him to lean back a bit and yelled, "Who the fuck do you think you are? Do you want me to FUCK YOU UP right here on this flight line? Get the fucking flush cart started, and we are going to flush these fucking engines!"

"Roger, Sir," he replied, almost falling over in fear.

We finally finished the work in an awkward silence, and he was on a flight with me two days later. I spoke with him and apologized for how I had handled it, and he apologized for his part as well. I learned something that day. You can't take back the words that have hurt people. Once they are out there, they are out there. You can, however, come back with better words, maybe words you should have used in the first place, and recover from the blunders you have made. This wouldn't be the last time I would have to do this in the deployment.

On November 1, I submitted my request for the aviation bonus. My earliest available retirement date would now be December 1, 2022, which meant three more years of service, potentially five more if fate allowed. In February 2020, I viewed the aviation dashboard, which provides a breakdown of the number of aviators holding various positions and operating in specific airframes. For the first time that I had ever seen it, Black Hawks had more MTPs than they did IPs. It was something like 79 percent to 74 percent, respectively.

I was talking with one of the IPs who arrived early for shift change and said, "They should send me to IPC to be a dual-tracked MTP/IP."

The IP was one of those boisterous loud talkers who said, "That would be awesome! Then, when we all get home, you could fly the initial progression flights that we all hate to fly, and we could do all the paperwork and make sure it's straight, which you would hate to do. It's win-win!"

At that exact moment, the CO walked in, saw the IP in his usual loud mode, and asked, "What's going on?"

The IP explained and the CO cut off the IP to say everything the IP had said to me about progression flights. We laughed it off, and I didn't think much of it until the next day.

I was on the day shift, and the IP and CO were on nights. As I arrived for the shift change, the CO said, "Chris, come digitally sign this paperwork." He had already filled out a 4186 or 4187, whichever number the paper was, for me to go to IPC. "There's a Mike Model IPC class starting in April. If we send you home in the next few weeks, you can have time to take some leave, spend a couple of weeks with your family, go to Rucker, become an IP, and you'll be back at Drum before we get home from deployment. You can progress as an IP and fly NVG resets with us before we go on leave. Then, you can keep flying folks while we are on leave. I will push it through with a statement like that to see if it works."

I signed the paperwork and thanked him, amazed and somewhat bewildered. He sent the email and left for the end of his shift, and I walked off, getting ready for my day, daydreaming about the possibilities of what was to come. I was planning far ahead now: finish my time at Drum, request to be stationed at Rucker for an instructor position, make W4, give two more years, drop retirement paperwork, retire as a twenty-five-and-a-half year W4, and maybe roll right into another flying job down there and stay in the Rucker area. Just like Kiowas, my dreams weren't meant to be.

Not even half an hour later, our BnAMO (battalion aviation material officer) called, "Chris, phone call in the CO's office." I heard. "Kruger." I stated after lifting the phone to my ear. "What's this email and paperwork for you to go to IPC?" He then spent the next fifteen minutes in a light diatribe about how I should focus on maintenance; dual tracking IP and MTP isn't a thing (even though it is). How would that work if I had a progression flight and the aircraft had an issue? Was it a maintenance flight, then? Did I get a new crew? Which took priority for me, instructing or maintenance? None of these were his problems or concerns, but since he made it his concern to tell me that and finish it with a rather blunt, "You should be focused on maintenance." I realized they would never let me branch out from my position, and I was done. I. was. fucking. DONE!

After that, I immediately called my wife, and even though it was very late for her due to the time difference, she picked up. I said, "Summer of 2022, we're out."

She said, "Wow! Something bad must have happened but all I needed to make plans after your retirement was the date of when it would be. I guess I can start putting together my resume!" She said with a reluctant chuckle. "So, what happened?" I told her everything, and she could hear my defeat, disgust, and anguish. I had given the Army nearly twenty years of my life. I had beaten up my body, damaged myself in ways I still may not know, carried crippling anxiety— straight up debilitating at times—and I was willing to give them more if they gave me one fucking thing. But no. I was done, and needless to say, my attitude turned to shit too.

And then there was COVID, and it was all about masks, distancing, reducing the number of passengers on aircraft and in the office, and ensuring roommates worked opposite shifts. We dealt with it all the best we could, just like the rest of the world. We continued business as usual as much as possible, and it didn't affect us significantly until we started coming home. I was rather fortunate on that front, though. My CO, a pretty cool dude, placed me on torch going home since first, I was on torch deploying; second, I now had seventy months of total deployed time, and third, he had another MTP. COVID affected the redeployment time-lines by delaying the relieving unit a month, so while we were all supposed to be going home in the first weeks of June 2020, only torch stayed on that timeline.

I arrived in Kuwait on June 4, 2020, to head home. There had been rumors that we would have to quarantine, but the details were unclear. The irony was that we were in the tran-sient barracks, where we all shared a common space with at least a hundred others. Bunk beds, wall lockers, everything had been previously used by someone else passing through, so the possibility of containing anything, especially an air-borne virus, was moot. I focused on completing the last two classes I needed for my associate degree, and even if we had to go into quarantine, I planned to use that time to finish them so I could go home with minimal interruptions. We had planned to go home a day later, but there was no word on quarantine back in the States.

Two days later, we got on a plane. It wasn't the regular garbage the Army gets contracted; it was a nice, brand-new Delta aircraft Before we got on board, there was a delay due to a mechanical issue they were trying to fix with the plane. While we were waiting, we finally got word that we would be going into quarantine in some unoccupied barracks upon our return home. Delayed and with discouraging informa-tion, I called my wife to tell her the news.

Understandably, she was pissed. She couldn't believe they would make such a decision at the last minute. "What's the difference if you are in some barracks or at home? Are any of us safer whether you are here but not home versus just at home?" I listened to her arguments about the obvious and stupid nature of the whole ordeal but managed to keep myself calm in the process.

I reminded her that at least I was coming home on time, not seven weeks later, like the rest of my company.

I reassured her that I would use the time to finish my classes and that by the time I got home, there wouldn't be anything pulling me away from them. Besides, now we were the only ones who knew, and we could tell the girls something else and surprise them with my arrival somehow. After the initial anger subsided, we agreed that this would be OK and, hopefully, I wasn't bringing COVID home. At least this way, we would know before I walk in the front door.

We finally boarded two hours later, and after a total of twenty-five hours on that aircraft, we arrived at Fort Drum and were whisked away with our baggage to a barracks where we would stay in quarantine for the next ten days. I used that time to complete my college classes and get some sleep. Soon, it was finally time to see my girls.

My wife and I decided to surprise our daughters with my arrival. Jenny would come to pick me up when we were allowed to leave at 8 a.m. She would drop me off a few houses down and around the corner from our house so that the girls wouldn't see me get out of the car or walk to the house. I would give her a few minutes to get back inside and distract the girls while I walked up and rang the doorbell. Since I wasn't home on the day I was supposed to be, we told the girls that we were all sent to Fort Riley for quarantine and would be home in two weeks, which on that day would have been two more days. We played it up on the phone calls, too; I was an hour behind them. It looked similar to Fort Drum, flat and green, with temperatures about the same, and the buildings looked the same because all Army buildings tend to look the same. They bought every last bit of it. Then the doorbell rang.

Jenny sent them to check and see if it was one of their friends at the door. Both girls were still in their PJs; I was in uniform. They opened the door, and I said, "Hey! Did you miss me?" Blank stares. They couldn't believe I was standing there. They didn't move in for hugs or say anything; they just stared at me. I was surprised by that, considering how they were when I left.

On my first aviation deployment, Charlize was two and a half years old, and Isabelle was two weeks old. Neither one of them remembers it, but it was hard on me. The next one, they were four and one and a half. They knew I was leaving but didn't fully understand that I wouldn't return for a long time. Also, they didn't remember that one. This time, they were almost nine and a little over seven. They knew what was going on, the potential danger I faced, that they may never see their daddy again, and they broke down and bawled, which caused me to do the same.

So there I was, back home, and blank stares until they realized it was real and smiles adorned their beautiful faces as they moved in for a hug. They hugged me tightly and didn't let go for a moment, nor did I motion for them to let go. I was home, and God willing, I wouldn't ever have to leave again.

Relieved to be home, having finished my associate's degree in general studies, with COVID-19 delays and associated complications in full swing, I was focused on preparing for retirement. Even though I knew it was two years away, I saved up as much leave as possible, ensuring all my medical records and physical issues were straightened out well before the VA physicals and getting a head start on the retirement processes.

What made matters worse was that I transitioned from having a great chain of command while deployed to one that was the polar opposite. From the top down, we had micro-managers who made those last couple of years excruciatingly miserable for me and for the majority of the unit too. It was so bad that I thought I would coin a new term, nano-manager, to describe someone who goes above

272 • CHRIS KRUGER

and beyond micro-managing, but it already exists in the Urban Dictionary. Add in COVID and having to meet electronically via multiple approved methods to chat virtually while simultaneously getting work done that requires collaboration with multiple people on one project. You get the idea; I was in for a rough two years.

I was less than two years out and had already started some classes I needed to complete to retire on time. The Army allowed retirees to start the SFL-TAP (Soldier for Life-Transition Assistance Program) two years from their retirement date, and I took advantage of it so I would have a reason to ditch work for the day and regularly.

After narrowly avoiding another trip to JRTC, a Bell's palsy flare up on the other side of my face, DES (Directorate of Evaluation and Standardization) coming to town while I was still on a down slip, and the unit was still on what I considered a downhill slope, my stresses combined to hit a new high. Did I man up and overcome? Not at all. I resorted to drinking again. I wasn't concerned about DES; I even passed their written exam well, considering there were recently some significant changes to chapter nine's emergency procedures. They even included some vehicle recognition and surface-to-air missile identification for good measure, which my prior infantry training had been helpful with.

It was a combination of the exhaustive nature of my physical, mental, and circumstantial stressors that I allowed to have control over me in a moment of weakness. So, I drank. Nearly three years of sobriety were tossed out the window. I initially maintained a good deal of control.

Months later, it got out so of hand again that I decided to call Behavioral Health, got scheduled to see the doc, and was back in therapy. It was recommended that I give medication therapy a try for both my behavioral health and sleep apnea issues, as I wasn't sleeping well either. I agreed. In May, I received all my medication and was rather excited at the possibility of it helping relieve my issues. I missed the part where I wasn't supposed to drink while taking these meds. The first night I took the sleep medication after a couple of drinks, I woke up with my face hurting and my wife on the phone with 911.

CHAPTER TWENTY

The combination of everything resulted in low blood pressure, and when I got up to go pee, I passed out over the toilet. My wife couldn't get me up as I crumpled straight down into a hugging-the-entire-toilet position. I woke a moment later, took a step back out of the bathroom, turned to my left, and as I was telling her, "I'm fine," I passed out again straight down on my face. The only reason I didn't lose a tooth, or worse, is that the CPAP mask was still on my face.

I came to a moment later. Jenny was frantic with 911 on the phone. I was just trying to crawl to the bed, climb back in it, and then I began to shiver. The EMTs came and took me to the hospital, where I stayed the rest of the night in the ER, having tests run that concluded the sleep meds were the culprit. A couple of drinks didn't help either.

The next day, I dumped the meds and decided that continuing in cognitive therapy would be the best, but I didn't wholly commit to the process. I went to counseling, but I was mainly just there. I told them everything I wanted them to know and left it at that.

Fast forward to August. My wife and I were seriously considering divorce. Understandably, she didn't want to deal with my BS. I was working on my retirement paperwork and was about a year from terminal leave. I had no future job prospects lined up; I didn't even know what I wanted to do, and work was about as stressful as it had ever been.

They now wanted me to fly flight lead on a four-ship Air Assault that I did *not* have the time to dedicate to the planning process with all the other maintenance issues that were on my plate. All this, and I had a psychological re-evaluation that same afternoon. I walked out to the hangar floor to catch a breath, looked to my left, and saw stairs that led up to the large hangar hoist and maintenance platform above and thought, *It's time.*

It's weird to think something so simple. *It's time.* I gave life all I could, the Army all I could, and this was it. My culminating life event would be to toss the last bits of whatever I had over a ledge for all to see. They would finally see what I was driven to by being overworked, underappreciated, overlooked, forced into an airframe, a track, a duty station, and anything else. All the good things came to me when I was in the infantry. I didn't have to fight for them, and I wasn't even in SF, SFOD, or SMU. Zodiacs and water train-ups? Here you go. HALO school, jumps, and other fun training? Gotchu' bro. In aviation, when I asked for a school or duty station, I got, "Go fuck yourself and do it dry." I walked up the stairs, along the platform, looked over the edge, and stared. I wasn't the first person I knew who would commit suicide in an Army hangar and I probably wouldn't be the last.

As I looked, I began to think. I was warming up to the idea. *How long will it take for someone to find me?* Probably within one minute, maybe two. Someone would come along and see me. *Faith-wise, what happens after suicide?* Is this an unpardonable sin? *I'm about to find out.* What does this say about my faith? What will people say after they find me? Was there a secret pool going on somewhere? Will someone win a bet? "He was bound to snap sooner or later; glad he didn't take anyone with him."

Staring, thinking, and working up the nerve to throw my body over, I stepped right up to the metal edge and placed the tip of my toes on it. I leaned over the railing and all I saw was the ground. It wasn't faith, it wasn't friends, and in all honesty, it wasn't even my family per se; it was fear that stopped me.

I had been a perfectionist to the point of fear all my life. In everything I had done, I held myself to the highest, and most likely, unachievable standard, and if I didn't reach it, I considered everything I did to be a failure. And that kept me from jumping; I feared that I would fail. I would live through it and be a paraplegic; my wife and kids would have to take care of me day-in and day-out for the rest of my life. There would be resentment toward me for the rest of their lives for being so selfish. They, nor would anyone else, ever forgive my selfishness and they'd hate me for what I did to them. Living through it would be worse for me and everyone else, so I walked away from the ledge, sat down, and cried.

That afternoon, I had my psychological reevaluation, and I went full in. I told the doc that mere hours prior, I had considered suicide. It wasn't the first time I considered it, which was true. When things got too hard, it was a common thought that I had, but until then, I hadn't taken the steps to accomplish the goal. Plenty of times, I had the means: weapons, helicopters, high buildings, whatever; but never had I taken those steps to be right there and ready to do it. I was 99 percent truthful with him that day, as I had only ever been about 90 percent truthful with psych docs in the past. I wasn't concerned about the down slip and not flying. I didn't care one iota about that anymore. I wanted to get better. I really wanted to get better. I didn't know it then, but the day prior would be my last flight, and as I always joked, it was a MTF. August 15, 2021 was the last time I touched a set of flight controls. I have no regrets about it; I just wish I had known it would be the last flight as I would have flown for a little bit longer.

Right around this time, I submitted my retirement paperwork. I was dropping it early, and I'm glad I did. Not only did we have a challenging chain of command to deal with, our S1 sucked! I had to go weekly, in person, and check the status of it. Even then, they routinely kept messing up the status of where it was. One day, they said it was in the office being checked for corrections; the next week, it was with the BN CO, and the following week, it was at BDE. In actuality, it never made the first step out of S1. It wasn't until the first week of November that I was finally contacted by the useless S1 OIC CPT that some changes needed to be made. I jumped on it, and since my retirement date was December 1, 2022, it needed to be sent out as soon as possible, I was in their office daily to ensure it happened.

All this while half of our company was going to Iraq, I was re-engaging with counseling, medication, and sobriety. I was then informed that, for some reason, I would be going to JRTC in January. I decided to go hands-off of all things maintenance, and when asked about anything, I would tell them, "I'm focusing on my retirement." Usually, I would get left alone after that. Usually.

New rounds of counseling and medications began with a new doctor, and this time, it was more productive. Come to find out, my time at Shank and being rocketed every day had fucked me up more than I thought. My personality traits, combined with years of never feeling adequate in aviation, or infantry for that matter, had produced anxiety, depression, and suicidal tendencies in me. This was complicated by the survivor's guilt I felt over all the friends I had lost during the years. There were abandonment issues as well, from when my physical security needs hadn't been met.

For example, you might recall that at Shank, the only thing we could do for safety was to place more sandbags around our beds to protect against incoming rockets. How sad is that? We are the strongest, most well-funded Army in the world, and the fix for incoming rockets is sandbags. Fucking sad. I worked with this doc for six months, and in that time, we went from meeting weekly to every other week to monthly at the end, and then he cut me loose on both the sessions and the meds. I was a few months from terminal leave, had kept my drinking mostly under control, and was doing OK mentally. Or so I thought.

In November 2021, the Krugers made it to Disney World for the first time. We hit four parks in five days and saw as much as we could in that time. The worst part of it was that we all finally got COVID. Having vaccines didn't matter. We got it.

We spent most of December in quarantine at our house, but the twice-daily phone calls from Army medical people and New York state healthcare officials were most frustrating. One day, I didn't hear my phone, so one of the healthcare groups called someone in the chain of command, and we were one step away from someone coming to the house to check on us. From then on, we decided we wouldn't bother getting checked if all we had were cold symptoms since that started the whole ordeal. It's ridiculous to think that we only had cold symptoms and were fine after a couple of days, but Jenny's Uncle died a few months later after a rather horrendous bout with it all. He was in his mid-sixties but in great shape, having been a bodybuilder and still a regular at the gym. The loss still saddens us, as he was a genuinely great guy.

I was hoping that my recent round of COVID would remove me from JRTC, but alas, on January 8, I was on my way to Louisiana one last time. I was supposed to be part of the command team or something stupid, so I only needed to take comfort items since I wouldn't be going to the field. We were in hardstand buildings with bathrooms, showers, and laundry. We had a couple of rental vehicles for the duration as we needed them to shuttle guys back and forth for shift work. Again, it was another complete waste of three weeks.

My job was to review AMRs (Air Mission Requests), inform the requesting unit of any necessary changes, and forward them to a major, a permanent party who worked at JRTC. I was a middleman for typos and missing information on paperwork. I read books, and when I was off shift, I hid some drinking.

I was on sleep meds again and one night, I had a tad too much to drink hanging out with some of the other guys and almost relived the ER visit from six months prior. I went to the bathroom down the hall and started to feel lightheaded. I stumbled down to my knees a couple of times in the hallway but was able to make my way back to my room and bed. As I collapsed into bed, breathing deeply and heavily, I decided I had better watch my consumption.

While there, I learned that my retirement paperwork had been accepted and approved at HRC. I would begin terminal leave on July 31 and officially retire on December 1, 2022.We were also able to close on the sale of our house in Fayetteville, which had been a continual thorn in our sides for the last twelve years or so. Things were looking pretty good, and of course, I would try to sabotage it with a bit of heavy drinking.

Once I returned from JRTC, I stopped going to work. I only shaved and put on a uniform when I needed to go in and check on some paperwork or email. Fortunately, I wasn't called in for much either, as mid-afternoon drinking became the norm. I kept it in check, mostly, as I was also transitioning off the depression medication. While I was still taking the sleep meds, I kept the drinking down but it wasn't doing me any favors.

We planned to retire to Clarksville, Tennessee, as we liked the area and were familiar with it. A series of discussions and prayers on the matter led us to consider Huntsville, Alabama, instead. When we decided on Huntsville, everything started falling into place quickly, and we took it as a sign. There was a realtor and a house that came quickly in a hot, albeit inflated, market that would work well for us. Job opportunities for both of us were endless, and the schools in the area we chose to move to were supposed to be the better ones. Jenny wanted to return to work, and we were both OK with my taking some time off to decompress and enjoy retirement while she played her hand at the job market. Everything was lining up, and we were in cruise control, ready for the next chapter for the Kruger Clan.

Jenny and the girls had to move to Alabama before I finished at Drum, so I flew down at the end of April and closed on the house. Some friends we knew there checked on it every other week as we needed mowing and had some things delivered to the house. The school calendars in New York and Alabama ended and started at different times, so the girls finished school in mid-June, came down to Alabama, received the HHGs, and started school on August 4. They only had about five weeks of summer break. I didn't have any leave to spare, but I took a chance and flew to Alabama for ten days to help set up the house before returning to finish outprocessing and retire.

Since everyone at Drum was on block leave at the beginning of July, I wouldn't be bothered as I had no appointments and wasn't flying anyway. We worked hard to set up the entire house and spent the rest of the time exploring the area and having fun before I headed back to Drum.

Of course, no one was around to celebrate since half of the company was deployed, and at that, the ones I would have wanted to celebrate with anyway. I only had a couple of appointments, and other than that, everything else I had to do would be at the end of July before I left, so I drank. I would get up, work out, eat some breakfast, and then drink vodka, or some of those nasty sparkling alcohol drinks, or whatever. I was staying with a buddy who was too busy to notice or maybe just didn't care.

One day, I'm sure I had alcohol poisoning and decided that this wasn't how I wanted to start my retirement. All that prompted me to do was to reset and recalibrate my drinking. Rules: nothing before noon, no more than a six-pack equivalent, buy what I need to make that happen, no bulk buys, *blah, blah, blah*. It worked for a time, but the train would jump the rails again.

At 10 a.m. on July 29, 2022, I left Clark Hall on Fort Drum, New York, for Huntsville, Alabama. It would take fifteen-and-a-half hours straight through. Columbus, Ohio, was the halfway point, and I told Jenny I would determine how I felt there. If I were OK, I would power through. If not, I would grab the cheapest motel, get some sleep, and roll out first thing in the morning.

It was just me in my Mazda 3. I would only stop to pee, get food or gas, and I'd be on my way. After a little over an hour, I passed through Syracuse, New York and one major city was down, with plenty to go. Then there was Rochester, Buffalo, and on I went into Pennsylvania. I passed through Erie and a small part of Pennsylvania into Ohio. Cleveland and Columbus, came next and I called my wife. I still had a good bit of daylight and I was feeling good, making good time, and driving on. At Cincinnati the sun started to set.

I crossed into Kentucky, hit Louisville, and I called Jenny again. I hadn't encountered any crazy traffic, not even a slowdown. I was feeling fine and going home.

"I've got about four and a half more hours, babe."

"Drive safe. See you soon," she replied.

It was a long stretch to Nashville, and I knew from previous drives that it was only about an hour to the turn-off for Huntsville after Nashville.

I finally stopped for gas about an hour from home. At this point, I hadn't gotten out of the car for about four hours. It was midnight at a busy gas station, and as I swung the door open to step out, I nearly collapsed but caught myself on the car door. My legs were somewhat numb, like when you've been sitting on the toilet too long, and my back was in severe pain that I overlooked from sitting and shifting my weight the whole time. I'm glad there weren't any cops around, as I do not doubt that I looked drunk for a moment.

After I grabbed the nozzle and put it in the gas tank, I stretched and walked the length of the car a time or two to look like a weary traveler, which I was, instead of a drunkard. Finishing up, I got in the car, shot a quick text to say, "See you in an hour," and I was off. I rolled into the driveway of our new life at 1:00 a.m. on the dot. Fifteen hours in a car, I was shaky, ready to shower and sleep.

We lived happily ever after, right?

I tried. I wanted to start a new chapter where I could be the guy who could buy some beer or liquor, have a couple of drinks on the weekend, and whatever I bought would last longer than a night. I was successful in that for about a week. Then, it was the same old thing: Get up, workout, have breakfast, do something that resembled work, day drink but not to the point of drunkenness, cook dinner with the family, go to bed, and repeat.

In September, my wife had some follow-up surgery from a previous one, and I decided that while she was recovering, I would remain sober. Then October hit, and we weren't jiving again; the reality of paychecks ending was starting to sink in. I had a pension and would have something for VA disability, but would it be enough? Did we plan enough like we thought? With all these uncertainties and issues in our marriage, I drank, and I drank hard.

Mid-October and everything came to a head. I was so drunk that Jenny was scared of what would happen that night and wasn't having any more of it. I was blackout drunk and don't remember the conversation at all. She does, and none of it was good on my half. The following morning, as I was nursing a rather horrendous hangover, she gave me the ultimatum: "Us, or alcohol. You choose."

CHAPTER TWENTY-ONE

I t was an easy choice. I re-engaged in my sobriety and have not looked back since. My faith, which I hadn't been living out, my marriage on the thinnest of ice, and my manhood and fatherhood under question, had all left my character in shambles. On top of that, I had reached 200 pounds, a weight I'd never been before, and the gain certainly wasn't in muscle mass. Being that out of shape also affected my sleep, which was already poor to begin with. Something had to change, and it needed to be holistic.

I had already determined that 2023 would be a year of change, and I take pride in saying it was just that. On January 2, 2023, I returned to the gym. I also decided to join our church's Bible in One Year Challenge. I joined a weekly men's Bible study group to reinforce the readings, enjoy additional Christian fellowship, and meet more church members for further encouragement in these endeavors. My wife and I attended counseling together, which greatly helped our marriage. I also started participating in weekly meetings at the Vet Center, which helped me appreciate the struggles I've faced and put them into perspective, allowing me to leave those issues behind and not let them drive my emotions or influence my decisions in the future.

That's the bottom line I want to emphasize. No matter your hardship, your mental struggle is very real. Could someone else have managed my twenty-two-plus years of experience better than I did? Sure. In fact, I could have done better myself. I stood on the brink, ready to end the suffering.

All I can say to anyone grappling with the challenges I've mentioned is this: someone cares! There is someone who would be worse off in this world without you here. No matter how you feel right now, someone needs you. Someone wants you in their life! Help exists out there. You may have to search for it, just as I did, but you will find it. Don't give up after just one attempt. Keep pushing until you find the freedom you seek.

I have embarked on a journey to transform myself and how I perceive my life, and this book was a part of that journey. I delved deep into my faith, sought mental and physical health, focused on family, and got external guidance. My wife will tell you that I am a different person now than when I started this journey, and I'm here to tell you that you can be a different person too. I've heard it said, "If you want the after picture, you can't keep living the before life!"

To that I add, don't give up the fight!

ACRONYM GLOSSARY

1SG: First Sergeant
AAR: After Action Review
ADSO: Active Duty Service Obligation
AFAST: Alternate Flight Aptitude Selection Test
AIT: Advanced Infantry Training
ALP: Air Load Planner
AMC: Air Mission Commander
AMOC: Aviation Maintenance Officers Course
AMR: Air Mission Request
AO: Area of Operations
APART: Annual Proficiency and Readiness Test
API: Armor Piercing Incendiary
ARS: Amphibious Reconnaissance School
ARTO: Assistant Radio Telephone Operator
ASEK: Aircrew Survival Egress Knife
ATC: Air Traffic Control
ATL: Assistant Team Leader
ATV: All-Terrain Vehicle
AWOAC: Advanced Warrant Officers Aviation Course
AWOL: Absent Without Leave
BAH: Basic Allowance for Housing
BCS: Basic Combat Skills
BDE: Brigade
BDU: Battle Dress Uniform
BH: Behavioral Health
BN: Battalion

BnAMO: Battalion Aviation Material Officer

BNCOC: Basic Non-Commissioned Officers Course

BN ORG: Battalion Organizational (Day)

BSO: Battalion Surgical Officer

BTLS: Basic Trauma Life Support

BUB: Battle Update Brief

BWS: Basic Warfighting Skills

CHU: Containerized Housing Unit

CO: Company Commander

CoC: Change of Command

CP: Command Post

CPAP: Continuous Positive Airway Pressure

CPL: Corporal

CPT: Captain

C-RAM: Counter Rocket Artillery and Mortar

CSM: Command Sergeant MajorCW2: Chief Warrant Officer 2

CW3: Chief Warrant Officer 3

CW4: Chief Warrant Officer 4

CW5: Chief Warrant Officer 5

DCU: Desert Camouflage Uniform

DES: Directorate of Evaluation and Standardization

DFAC: Dining Facilities

DOR: Drop on Request

DZ: Drop Zone

EIB: Expert Infantryman Badge

EMT: Emergency Medical Technician

ENDEX: End of Exercise

ENT: Ear, Nose, and Throat

EOD: Explosive Ordinance Disposal

ERB: Enlisted Record Brief

E-3: Private First Class (PFC)

E-4: Specialist/Corporal (SPC/CPL)

E-5: Sergeant (SGT)

E-6: Staff Sergeant (SSG)
E-7: Sergeant First Class (SFC)
E-8: Master Sergeant (MSG)
E-9: Sergeant Major (SGM)
FARP: Forward Aerial Resupply Point
FL: Flight Lead
FLOT: Forward Line of Troops
FOB: Forward Operating Base
FRIES: Fast Rope Insertion Extraction System
GBU: Guided Bomb Unit
GPS: Global Positioning System
GSAB: General Support Aviation Battalion
HALO: High Altitude, Low Opening
HAMET: High Altitude Mountainous Environment Training
HHG: Household Goods
HLZ: Helicopter Landing Zone
HME: Home Made Explosive
Humvee: Highly Mobile Multi-Wheeled Vehicle or Humvee
HQ: Headquarters
HRC: Human Resources Command
HS: Hide Site
HUMINT: Human Intelligence
HVAC: Heating Vacuuming and Air Conditioning
HVT: High Value Target
IBP: Iraqi Border Patrol
IED: Improvised Explosive Device
IERW: Initial Entry Rotary Wing
IFR: Instrument Flight Rules
ILS: Instrument Landing System
IP: Instructor Pilot
IPC: Instructor Pilot Course
IR: Infrared
J-LIST: Joint Lightweight Integrated Suit Technology
JM: Jumpmaster

JMPI: JumpMaster Personnel Inspection
JRTC: Joint Readiness Training Center
JTAC: Joint Tactical Air Controller
KBR: Kellogg-Brown and Root
KIA: Killed in Action
KMC: Kilauea Military Camp
LAO: Local Area Orientation
LMTV: Light Moving Tactical Vehicle
LRSC: Long-Range Surveillance Company
LT: Lieutenant
LZ: Landing Zone
MALE: Maneuver, Abort criteria, Limitations, Emergency
    actions
ME: Maintenance Examiner
Medevac: Medical Evacuation
MGRS: Military Grid Reference System
MI: Military Intelligence
MILES: Multiple Integrated Laser Engagement System
MKT: Military Kitchen Trailer
MOPP: Mission-Oriented Protective Posture
MP: Military Police
MRAP: Mine Resistant Ambush Protected
MRE: Meal, Ready-to-Eat
MTF: Maintenance Test Flight
MTP: Maintenance Test Pilot
MTPC: Maintenance Test Pilot Course
NBC: Nuclear, Biological, Chemical
NCO: Non-Commissioned Officer
NOE: Nap of the Earth
NTC: Northern Training Center
NVG: Night Vision Goggle
OBJ: Objective
OC-T: Observer/Controller-Trainer
OPFOR: Opposing Forces

OPS: Operations
OPTEMPO: Operational Tempo
O1: Second Lieutenant
O2: First Lieutenant
O3: Captain (CPT)
O4: Major (MAJ)
O5: Lieutenant Colonel (LTC)
O6: Colonel (COL)
PC: Pilot in Command
PC: Production Control
PCL: Power Control Lever
PCNCOIC: Production Control NCO in Charge
PCOIC: Production Control Officer in Charge
PCS: Permanent Change of Station
PI: Pilot
PID: Positive Identification
PLDC: Primary Leadership Development Course
PLF: Parachute Landing Fall
PLT: Platoon
POG: People Other than Grunt
POW: Prisoner of War
PPW: Pacific Pathways
PRC: Pre-Ranger Course
PSG: Platoon Sergeant
PT: Physical Training
PTSD: Post-Traumatic Stress Disorder
PWAC: Practical Work inside the Aircraft
PX: Post Exchange
PZ: Pickup Zone
QC: Quality Control
QCNCOIC: Quality Control NCO in Charge
QCOIC: Quality Control Officer in Charge
QRF: Quick Reaction Force
RC: Regional Command

RCOP: Risk Common Operational Picture
RI: Ranger Instructor
RIMPAC: Rim of the Pacific
RIP: Ranger Indoctrination Program
RLO: Real Life Officer
RL3,2,1: Flight Readiness Level
RLTW: Rangers Lead The Way
ROE: Rules of Engagement
ROZ: Restricted Operating Zone
RP Rally Point
RPG: Rocket Propelled Grenade
RSLC: Reconnaissance and Surveillance Leaders Course
RTB: Return to Base
RTO: Radio Telephone Operator
SAT: Satellite
SAW: Squad Automatic Weapon
SERE: Survive, Escape, Resist, Evade
SF: Special Forces
SFOD: Special Forces Operational Detachment
SGM: Sergeant Major
SGT: Sergeant
SMU: Special Mission Unit
SO: Scout Observer
SOP: Standing Operating Procedure
SP: Standardization Pilot
SPC: Specialist
SPIES: Special Insertion Extraction System
SS: Surveillance Site
SSG: Staff Sergeant
SSO: Senior Scout Observer
S-Vest: Suicide Vest
S1: Administrative Office
S2: Intelligence
TBI: Traumatic Brain Injury

TC: Truck Commander

TF: Task Force

TL: Team Leader

TMC: Troop Medical Clinic

TOC: Tactical Operations Center

UAS: Unmanned Aerial System

UMO: Unit Movement Officer

UPL: Unit Prevention Leader

VA: Veterans Affairs

VFW: Veterans of Foreign Wars

VOR: Very High Frequency Omnidirectional Range Station

WO: Warrant Officer

WOCS: Warrant Officer Candidate School

WO1: Warrant Officer

1SG: First Sergeant

ACKNOWLEDGMENTS

Although this book took a considerable amount of time to write, it would not have been possible without the support of many people.

First and foremost, I would like to thank my wife. Without her, none of this would have been possible. She encouraged me to share my story, and when I hesitated, she urged me to move forward. When the emotions of writing about my past–the good, the bad, and everything in between–which she already knew so much about, overwhelmed me, she encouraged me even more. When I wanted to toss the keyboard and give up, she stood by me with understanding and once again urged me to see it through. Without her, I would not have met the people who are making this book a reality. Thank you, my dear wife; you've been there at my best and witnessed me at my worst, and we're still together, lifting each other up.

Thank you, Brunella Costagliola of The Military Editor® Agency, LLC. Your guidance and encouragement during the editing process have been invaluable to me. Whenever I felt like quitting for various reasons, you helped me recalibrate my thinking and get back on track. You also helped me achieve my highest potential in writing as we created a more enjoyable and engaging story. I look forward to seeing how this story impacts others.

I want to thank Annette Whittenberger. I read her book *The Wall Between Two Lives* about a year too late. We share similar stories under different situations and circumstances. Your encouragement to "make your mess your message" has inspired me to focus on ensuring my message reaches more people.

I would like to thank my family for their continued support in all the endeavors life has presented. I know you would all prefer for me to live closer to home. It hasn't been easy being apart for over twenty years. Still, I appreciate you setting that aside and encouraging me from a distance. If I never learn anything more than the fact that you are proud of me, my accomplishments, and everything I've become, that will be enough. I love you, and I'm so grateful for your support throughout my life.

To all my friends and former coworkers with whom I shared the rough draft, thank you for your encouragement. Your feedback has motivated me to continue pursuing this book, with the hope that it will resonate with people and encourage them to seek the help they may need.

I have to thank the military in general, without which I would not have had these awesome and horrible experiences that have shaped who I am. I am grateful for most of the experiences and people who have come into my life.

Finally, I thank God for bringing me to it and for keeping me through it all. My faith has grown and wavered throughout the many years of military service, but God has been there to see me through. Now, I get to share the experience and encourage others on their journeys, too.

ABOUT THE AUTHOR

C hris Kruger grew up just outside Spokane, Washington, and joined the Army before turning 21, driven by a sense of purpose and a thirst for challenge.

Over the next 22 years, he forged a diverse and demanding military career—starting as an infantryman and eventually becoming a Black Hawk maintenance test pilot. His journey took him from basic training and US Army Airborne School at Fort Benning to Fort Bragg, North Carolina, where he completed an impressive lineup of elite training programs, including Amphibious Reconnaissance School, Ranger School, HALO and HALO Jumpmaster, and flight school, among others.

Along the way, he earned the Expert and Combat Infantryman Badges, a Bronze Star, a Meritorious Service Medal, and several other honors.

Chris deployed to Iraq, Afghanistan, and the Far East, spending more than five and a half years overseas—experiences that deeply shaped his outlook on leadership, resilience, and life.

After retiring, he turned to writing as a way to unpack and give meaning to those intense years. He writes from a place of survival, strength, and connection. What began as personal reflection quickly became a mission to reach others walking similar paths.

Today, Chris lives in Huntsville, Alabama, with his wife, Genevieve, and their two daughters, Charlize and Isabelle. Together, Chris and Genevieve are active in their local church and are committed to encouraging others to overcome life's hardships, expanding their reach through writing and Genevieve's forthcoming podcast aimed at inspiring healing, faith, and perseverance.

www.ingramcontent.com/pod-product-compliance
Lightning Source LLC
Chambersburg PA
CBHW032148080426
42735CB00008B/632